A Practical

Introduction

to C

Paul Jarvis

Centre for Computing Services
Imperial College of Science, Technology and Medicine

Oxford   New York   Tokyo
OXFORD UNIVERSITY PRESS
1994

Oxford University Press, Walton Street, Oxford OX2 6DP

Oxford New York Toronto
Delhi Bombay Calcutta Madras Karachi
Kuala Lumpur Singapore Hong Kong Tokyo
Nairobi Dar es Salaam Cape Town
Melbourne Auckland Madrid

and associated companies in
Berlin Ibadan

Oxford is a trade mark of Oxford University Press

Published in the United States
by Oxford University Press Inc., New York

A catalogue record for this book is available from the British Library

Library of Congress Cataloging in Publication Data
Jarvis, Paul.
A practical introduction to C / Paul Jarvis.
1. C (Computer program language)   I. Title.
QA76.73.C15J38   1994   005.13'3–dc20   93-45532
ISBN 0 19 853846 4 (Pbk)
ISBN 0 19 853847 2 (Hbk)

Prepared by the author using Xerox Ventura Publisher
Typeset at the University of London Computer Centre

Printed in Great Britain by
Bookcraft (Bath) Ltd
Midsomer Norton, Avon

Preface

As the title suggests, this text is an introduction to using the C programming language. The idea is to provide a concise summary of the principal features of the language to enable the reader to use C to solve real problems. The reader is assumed to have some knowledge of programming in any language.

This book is an extension of a course presented by the author to both undergraduates and graduates at Imperial College, University of London. The course is usually given over a period of three consecutive days and contains an even split between lectures and practical sessions, with the last day spent entirely on projects.

This text follows a similar format: the first part of the text is a tutorial with exercises after each topic and the second part of the text comprises a number of projects that are designed to be stimulating and to bring together the techniques covered previously. Answers to both the exercises and the projects are given.

The more adventurous projects require access to an IBM/PC (or clone) as certain hardware specific features are used. Also, for the imaging project, an EGA display (or better) is required.

This text is designed to introduce C using a practical approach, and does not attempt to be a definitive statement of the language, as this is really the preserve of the reference manuals.

The C language was developed by Brian Kernighan and Denis Ritchie to provide a suitable language for their use. Subsequently C was adopted by others working in widely differing fields, and was not found lacking. This gain in popularity, together with the close association with the Unix operating system, resulted in C becoming one of the major programming languages.

As C became more widespread the need for a standard definition grew. The American National Standards Institute (ANSI) accepted the challenge and released their 'ANSI C' standard in Autumn 1989. Subsequently the International Standards Organization (ISO), and the British Standards Institute (BSI) have adopted the ANSI definition. For historical reasons the term 'ANSI C' is usually seen, rather than 'ISO C'.

This text uses the ANSI standard throughout. Where there are significant differences from the original definition then these are highlighted using vertical lines in the margin. It is a good idea to know both forms, but new programs should always follow the ANSI standard.

To Janet

Contents

1

Getting started

Aims and objectives

The aims of the chapter are as follows:

- to enter and compile a small C program;
- to practise editing (i.e. changing) the program;
- to see how errors are indicated by the compiler.

The task

The first steps of any new project are always the hardest. Learning to drive a car is a typical example where the simple instruction 'Put the car into neutral' requires knowledge of both gears and clutch, and how they are controlled. This is also true for running your first C program in that, before considering the language itself, one needs to know how to enter programs, compile them, and run them. By far the best way would be for another knowledgeable person to demonstrate it.

Entering programs

In general there are two ways in which any type of programming can be done. One is the 'environment' method, where a program is run that provides the mechanism to enter, compile, run, and debug programs. This is becoming more popular as mouse devices become common and menu/window–driven interfaces are the vogue. The Microsoft Quick C and Borland Turbo C products fall into this category.

The other method, most commonly demonstrated on Unix platforms, is where program entering, compiling, and debugging are all done as separate stages. There may be a choice of several different editors, personal taste dictating the one to use. Often there is more control over the compilation options and again a choice of debugging tools. Microsoft C fits into this category. Currently this method gives the greatest flexibility, although the environment method is often found easier for beginners.

All the examples in this text have been tested using both Microsoft C version 5.1 and Microsoft Quick C version 2.50. Minor differences may be found when using other products but this should be in the method of use, and not in the C language itself.

A first program

To start, enter the program listed in Fig. 1.1 exactly as shown. All the punctuation is required. The example is a standard initial program in all C texts and was first used in *The C Programming Language* by Kernighan and Ritchie. As this was the initial definition of the C language the author feels he is following in good footsteps.

Fig. 1.1   A simple C program

```
#include <stdio.h>
main()
{
   printf("Hello World\n");
}
```

When the above program is compiled there should be a warning message printed saying something about there being no return value. If no such message is printed consult your compiler manuals to see if such warnings are suppressed by default. If so take the required steps to get these warnings printed. More often than not a warning message indicates an error; however, in this case the message can be ignored. We will get rid of it later when functions are discussed in detail.

Having entered and successfully run the program, try changing it and induce a few errors. Make minor changes like deleting the semicolon (;) or one or more of the brackets. This is constructive as it will give a feeling for what will happen with real programs.

2

Grammar and vocabulary

Aims and objectives

The aims of this chapter are to:
- introduce the general structure of a C program;
- define global objects;
- show the structure of a function;
- mention the existence of the standard library;
- introduce the idea of a pre-processor;
- define and suggest how comments are incorporated.

Why is it needed?

The title 'Grammar and vocabulary' must be the biggest turn off so far. Unfortunately with any language, be it natural or computer, the basic building blocks must be mastered before vast constructions can be undertaken. One of the attractive points of the C language is that there are only a few basic rules to learn. There are no irregular verbs, the bane of all school children.

Program structure

C is a free format language. In other words the compiler ignores spaces, tabs and newline characters (collectively termed white space). White space characters are not permitted within names, and newline characters are not permitted in string constants. Both terms will be described in due course. It would be possible to write a program entirely on one line, at least as far as the compiler was concerned. This is not recommended as it is far easier to read a well laid out program, with indentation providing a visual clue of how parts of the program are linked. This will be shown further in the chapters covering control statements (e.g. `for`, `while`, `if`).

Global objects

A C program is composed of global objects. These either contain data or executable code (i.e. variables or functions). Functions can also contain data but this is local to the function and is not accessible outside the function. Functions cannot be defined within functions. One (and only one) function in a program must be called `main`. This is where the program starts executing. Now we have the general structure, let's go in for a bit more detail.

Function structure

A function has a name, optionally some parameters, and a body. The body consists of zero or more statements enclosed within braces (`{` and `}`). While this definition is not complete, it will suffice until the subject is covered more fully in Chapters 14 and 19.

```
name (optional  parameters)
{
    statement(s)
}
```

A group of statements contained within braces is known as a compound statement and, apart from function definitions, single statements and compound statements maybe used interchangeably.

Structure by example

Looking at the example program from the previous chapter (Fig. 1.1) it can be seen that—ignoring the first line—it is a function definition. The function name is `main`, there are no parameters (if there were they would be within the parentheses) and the body of the function contains a single statement that begins `printf`. Individual statements end with a semi-colon (`;`). It is amazing how many error messages can be generated when compiling a program with a missing semicolon. Statements can be continued over several lines if required, but lines must not be split in the middle of a name or, as will be described later, a character string.

The example program consists of a single function definition. As is required, this function is called main. This function consists of a single `printf` statement that is in fact a function call. In C there are no input or output statements (the equivalent of Fortran's `read` and `write`). Instead a standard library of functions is supplied with each compiler to provide these facilities. This has the advantage of simplifying the C language and also offers a greater range of features. This standard library will be covered in some detail throughout the text. To use a function one specifies the function name followed by any arguments (if required) surrounded by parentheses. Again, the subject of function use is covered fully in Chapters 14 and 19.

Pre-processor

The last section specified the rule that statements end with a semicolon. However, what about the very first line shown in Fig. 1.1, which does not end this way? The answer is that this is not a C statement; rather, it is a directive to a pre-processor through which all C source is passed before being compiled. This pre-processor is described in Chapter 18, but for now always include the following as the first line in every program written.

```
#include <stdio.h>
```

Comments

When trying to help students with programming problems it is amazing to see how seldom comments are used. Often there is nothing to indicate what a program is designed to do. Comments are important; you may know what you mean now, but will you next year when the program needs changing? In C, comments can be entered anywhere a space, tab, or newline character is valid. The comment is delimited by using the character pairs /* to start and */ to end. Comments can continue onto subsequent lines as required. As far as the compiler is concerned a comment, whatever its length, is processed as if it were a single space.

Do not nest comments. In other words, one comment cannot contain another one. It is important to remember this when commenting out a section of code while debugging a program. The following syntax is incorrect:

```
/*  This is an /* illegal */ comment syntax  */
```

The program shown in the previous chapter should really have been something like this:

Fig. 2.1 A better first program

```
/*  Program to print 'Hello World'    */
/*  P. Jarvis.           24/11/1990   */

#include <stdio.h>
main()
{
   printf("Hello World\n");
}
```

Suggested program header

With any significant program it is worth having a block of comments listing the purpose of the program, method of use, the author(s), and a brief development history. It is also important to document known limitations and bugs. Admitting a program has bugs is not a sign of weakness; rather, it is a sign of being realistic. Figure 2.2 gives a typical program header.

Fig. 2.2   Typical program header

```
/*   sorti - Program to sort numbers          */

/*   P. Jarvis.            15th December 1990  */

/*   Version 1.1                              */
/*           Extended to 1000 numbers.        */
/*           These can be read from a file    */
/*           that is specified as a para-     */
/*           meter. Uses a bubble sort.       */

/*   Version 1.0                              */
/*           Sorts numbers from standard      */
/*           input. Must be valid integers    */
/*           and no more than 100 values.     */
/*           Program aborts on invalid        */
/*           input.                           */
```

Summary

- Programs consist of functions and data.

- Functions have a name, optional parameters, and a compound statement.

- A compound statement is one or more statements enclosed within braces ({ and }). There should not be a semicolon after the closing brace.

- One, and only one, function must have the name main. This is where the program begins when it is run.

- C is a free format language. This means that 'white space' characters (i.e. spaces, tabs, and new lines) are effectively ignored by the compiler. Exceptions to this are that names must not contain such characters and character strings must not contain newline characters. Names and character strings have yet to be defined.

- Comments are a necessary part of all programs. A comment must be enclosed within the character sequences '/*' and '*/'. Comments must not be nested, in other words one comment cannot contain another.

Exercises

1 Referring to Fig. 2.1, identify the following parts of the program:
Comments
Function definition
Pre-processor directive
Statement

3

Data types

Aims and objectives

The aims of this chapter are to:

- introduce how numbers are stored;
- define octal, decimal, and hexadecimal number bases;
- specify the four basic data types of C;
- highlight the different ranges these data types may have.

Overview

Computers work in terms of bits, bytes, and words, whereas people use sensible terms like 'numbers'. To communicate properly, one has to modify one's concept of numbers. Integers, or whole numbers, are treated differently from real numbers (those which can have a fractional part). Thus although most people would accept that 1 is identical to 1.0, a computer would disagree.

The difference arises from how numbers are stored. Integer numbers, which can never have a fractional part, are stored in simple binary. Table 3.1 lists the binary and decimal equivalents of the first sixteen integers. The octal (base 8) and hexadecimal (base 16) equivalents are also shown as these will be used in later projects.

Number bases

For those not familiar with different number bases, here is a quick summary. The decimal number system uses both the position and the digit to determine the value of the digit. For example 2 means two while 20 means twenty. Each column position is valued as ten times the value of the column to its right. Thus the number 123 is $1 \times 100 + 2 \times 10 + 3$.

Octal

Octal works in the same way but each column is worth eight times the column to its right. Thus 123 octal is $1 \times 64 + 2 \times 8 + 3$ (eighty three in decimal). The largest digit permitted in octal numbers is 7, which is one less than the multiplying factor between columns.

Hexadecimal

Hexadecimal values each column as sixteen times the one to its right, thus sixteen different digits are required. 0 to 9 count for the first ten digits and, by convention, A to F are used for the remainder (either upper or lower case). The number 123 in hexadecimal is $1 \times 256 + 2 \times 16 + 3$ (two hundred and ninety one in decimal). The value thirty one decimal is written as 1F in hexadecimal (or hex) notation.

Table 3.1 Number bases

Decimal	Binary	Octal	Hex
0	0000	00	0
1	0001	01	1
2	0010	02	2
3	0011	03	3
4	0100	04	4
5	0101	05	5
6	0110	06	6
7	0111	07	7
8	1000	10	8
9	1001	11	9
10	1010	12	A
11	1011	13	B
12	1100	14	C
13	1101	15	D
14	1110	16	E
15	1111	17	F

Internal format of integer numbers

Table 3.1 shows the internal binary format of the first sixteen integers. Each bit position is valued as two times the column to its right and the only permitted digits are zero and one. Thus the binary number 101 is $1 \times 4 + 0 \times 2 + 1$ (i.e. five). This works for all positive numbers. For negative numbers there has to be a convention for their representation. The general rule is that if the left–most digit of an integral number is set to a one, then the number is negative. There are still two common ways to represent the numbers: ones complement and twos complement form.

To form the ones complement simply take the representation of the positive number and convert every one to a zero and every zero to a one. For example the number five is 00000101 while minus five is 11111010. Note that there are two representations for zero. There is a plus zero (i.e. 00000000) and a minus zero (i.e. 11111111).

The twos complement of a number is formed by inverting the individual bits (as for the ones complement) but then adding one. Thus minus five in twos complement form is 11111011. This form is more common (and is used on personal computers), and it has only one representation for zero.

Where the ANSI standard specifies a permitted range of values then no assumption on the internal representation is made. Thus, for example, eight bit quantities are specified as being able to hold numbers within the range of −127 to +127. If the machine used ones complement form this would be correct, but for a twos complement machine the actual range would be −128 to +127.

Real numbers

Real numbers (also known as floating point numbers) are stored in an exponential format where there is a mantissa and exponent. For example the number 123.4 could be written as 0.1234×10^3 (where 0.1234 is the mantissa and 3 is the exponent). For ease of computation (by the computer that is) the exponent usually works in powers of two rather than ten. Table 3.2 lists some real numbers together with their hexadecimal representation as used by the Microsoft C compiler.

Table 3.2   Typical real number representation

Decimal	Representation (base 16)
−1.0	BF800000
1.0	3F800000
2.0	40000000
4.0	40800000
7.5	40F00000
10.0	41200000
100.0	42C80000

Internal format of real numbers

For those interested in how real numbers are stored internally (and this understanding is not a pre-requisite to programming in C) the following describes how such numbers are generally stored on an IBM PC or equivalent. Where a number of bits is specified, it is assumed that the real number occupies a total of four bytes (32 bits).

Each number contains three parts: the sign, exponent, and mantissa. The exponent is base two rather than the normal ten. The value of a number is given by:

```
mantissa × 2^exponent
```

Sign

The left–most bit indicates the sign of the number. If this bit is a one then the number is negative. Compare the values for 1.0 and −1.0 in Table 3.2 and verify the difference is only in this bit.

Exponent

The next eight bits are the power of two exponent. The value is offset by 0x7F. In other words a value of 0x7F means 2^0 while a value of 0x80 means 2^1. Again by referring to Table 3.2 and comparing the entries for 1.0, 2.0, and 4.0 one can see the exponent incrementing each time. If in doubt, write out the hex digits in binary form (Table 3.1 might help here) and divide out the various sections, first bit sign, next eight bits exponent.

Mantissa

The remaining 23 bits are the mantissa. Although only 23 bits are specified, there is an implied 24th bit to the left of these 23 bits and this extra bit is always set. The extra bit is assigned the meaning 1.0, the bit to its right is half that, that is, 0.5, the bit to its right is then 0.25, and so on. The value of the mantissa is therefore the sum of the set bits.

Verify the value for 10.0 given in Table 3.2. The mantissa is 1.0 + 0.25, and the exponent is 0x82 − 0x7F (i.e. 3), giving a value of 1.25×2^3 (i.e. 10.0).

Date types in C

There are four fundamental types of data used in C. These are `char`, `int`, `float`, and `double`.

Character

The first (`char`) is defined as being large enough to contain any value which represents a character on the computer being used. Most computers use the American Standard Code for Information Interchange (ASCII), where, for example, the letter A is represented by the value 65. ASCII is a seven–bit code and thus chars are usually only one eight–bit byte; however, this is implementation dependent. On an IBM/PC the ASCII character set is used plus an IBM extension which creates a 256 character range. chars therefore require eight bits and are stored as one byte. Appendix A lists the ASCII standard and also includes the IBM extensions.

Integer

The type int is used to contain integer numbers, that is, those with no fractional part (e.g. 23, 175, −42). The size of an int is at least 16 bits so a number in the range −32 767 to +32 767 can be stored. The two qualifiers short and long can also be used with int, a short int being at least 16 bits and a long int at least 32 bits. When using the short and long qualifiers, the associated int can be omitted if preferred.

Note, the standard states that the size of short is less than, or equal to, that of int. Similarly the size of int is less than, or equal to, that of long. The standard also states that int is at least 16 bits and long is at least 32 bits. Otherwise the exact sizes are implementation dependent.

Unsigned prefix

Both char and int may be prefixed by the qualifier unsigned. If an int is defined as being 16 bits it can represent numbers in the range −32767 to +32767. However an unsigned int can represent 0 to 65535. unsigned chars will be used frequently in the exercises to hold values in the range 0 to 255.

Real numbers

The two remaining data types are float and double. Both represent numbers with a fractional part (e.g. 1.23, 3.14159), just the accuracy and maximum values vary. The qualifier long can be applied to double. This then gives three sizes of floating–point number. Their sizes are implementation dependent but the maximum range and accuracy of float is less than, or equal to, that of double. This, in turn, has a resolution and range which is less than, or equal to, that of long double.

The variable type long double was introduced by the ANSI standard. Previously this feature was not available.

Table 3.3 lists all the valid data types as used by the Microsoft compilers on the IBM/PC.

Table 3.3 Microsoft C data sizes

Type	Bytes	Range
char	1	−128 to +127
unsigned char	1	0 to 255
short int	2	−32,768 to +32,767
int	2	−32,768 to +32,767
long int	4	-2^{31} to $+2^{31}-1$
unsigned short	2	0 to 65,535
unsigned int	2	0 to 65,535
unsigned long	4	0 to $+2^{32}-1$
float	4	$\pm1.2E^{-38}$ to $3.4E^{+38}$
double	8	$\pm2.2E^{-308}$ to $1.8E^{+308}$
long double	16	$\pm3.4E^{-4932}$ to $1.2E^{+4932}$

Summary

- Computers differentiate between whole numbers (integers) and rational numbers (floating–point or real numbers).

- C has four basic data types. These are: char, int, float, and double. The unsigned keyword may be applied to char and int. The keyword short may be applied to int and the keyword long may be applied to both int and double. This creates a total of eleven possible data types.

- The absolute resolution of these types (i.e. how large a number they can store and, for real numbers, their accuracy) is not specified by the standard, only minimum values are given. Therefore the sizes of these data types will vary from one machine to another.

Exercises

1 List the four basic data types used in C.

2 Calculate the values of the following numbers if they are evaluated as octal, decimal, and hexadecimal.
17
100
321

3 If on a given system an `int` can take the values from −32 768 to +32 767, then what would be the range for `unsigned int`?

4 Using the floating point representation described above what is the value of 0x3FF80000, and what is the internal form of 6.0 (1.5×2^2)?

4

Constants

Aims and objectives

The aims of this chapter are to:
- show how to define constants for each data type;
- introduce escape characters in character constants;
- define string constants.

Overview

Given that there are four data types the next step is to define how to specify constants for each. Obviously a constant of type int must not contain a decimal point, but one of type float must. However, is a short int zero the same as a long int zero? Unfortunately not.

Integer constants

An int constant can be represented by a sequence of numeric digits which do not include a decimal point. If the number is too large for an int then it is automatically assumed to be of type long. The suffixes U and L (either upper or lower case) mean unsigned and long respectively. (It is probably a good idea to avoid using lower case L as a suffix as this can easily be mistaken for the number one.) Thus 0 and 0L on an IBM/PC are different: the first is a 16 bit zero value, the second is a 32 bit zero value.

There are a couple of points which might cause one to come to grief. When writing operating systems and the like—one of the first uses of the C language—octal (base 8) and hexadecimal (base 16) numbers are frequently used. To define a number in octal, precede the value with a zero. Thus beware of putting leading zeros in constants when not required. So 11 is decimal eleven but 011 is decimal nine. This point is often forgotten at first. To specify a hexadecimal number precede the value with 0x (zero followed by the letter x). Thus 0x11 is seventeen decimal. We will be using octal numbers, and some hexadecimal numbers, in the examples and the projects in the second part of the text. The following example summarizes the different forms.

```
   11              eleven  decimal
  011              nine  decimal
0x11              seventeen  decimal
```

Real constants

There are three sizes of real constants. These are `float`, `double`, and `long double`. All three represent real numbers: only their permitted size varies. Note that `long double` is available only when using ANSI C compilers—it was not available in the original definition.

float

Constants of type `float` are specified by numbers containing a decimal point and followed by the suffix `F` (or `f`). The suffix is required as without it the constant will be assumed to be of type `double`. This is not as silly as it may sound when one realises that usually arithmetic is done using `doubles`: `floats` are only used to save space and when precision is not too important.

double

The standard maths library functions all use variables of type `double` and the original definition of C specified that all non-integer arithmetic was done using type `double`. Variables of type `float` were converted to `double` and the result converted back to float automatically. For this reason, constants containing a decimal point and no other suffix are taken to be of type `double`.

long double

A constant containing a decimal point and having the suffix `L` (or `l`) indicates a `long double` constant. It is probably better to use the upper case suffix, as this will avoid confusion between the letter `l` and the number `1`. Remember that this type may not be available on pre-ANSI compilers.

Character constants

A `char` constant can be specified as an integer value within the legal range, or by specifying a single character between single quotes (e.g. `'A'`). The second case means 'the value which represents the letter A'. It is far better to use this form than specifying the numeric equivalent as it is clearer and is independent of character set (not everyone uses ASCII).

Non-printing characters

Sometimes it is difficult to enter a character because it has no visible representation. The backspace character is probably a good example of this. Simply entering a backspace character is not possible because pressing the backspace key will result in the previous character being deleted. To allow for this, and other special characters, C uses a character pair to represent a single special character. The first character is a backslash (\) and this is followed by a key letter representing a single special character. For example \b represents the backspace character.

If the key letter is not valid then the backslash is ignored. Table 4.1 lists the valid codes. Note that the key letter x requires a two–digit hexadecimal number following, while a numeric character after the backslash indicates an octal value follows. This value can be up to three digits long.

Table 4.1   Special characters

Character	Meaning
\a	bell (audible alarm)
\b	backspace
\f	formfeed
\n	newline
\r	carriage return
\t	tab
\v	vertical tab
\x*nn*	hexadecimal value
nnn	octal value
\'	single quote
\"	double quote
\?	question mark
\\	backslash

From the above one can now see that—on a system using the ASCII character set—there are at least four different ways of defining the character constant representing the letter A. These are:

    65   'A'   '\x41'   '\101'

String constants

There is one further type of constant which is the character string, or string constant. We have yet to define a variable type which can hold such data, but this will be described in due course.

A string constant is a series of characters delimited by double quotation marks. The backslash escape character can be used in the same way as with character constants. The string constant contains one more character than is visible. A zero value (i.e. the character '\0') is placed after the last character to act as a delimiter. As a string is not permitted to contain a zero value, this additional byte indicates the end of the string.

We have already seen an example of a string constant in the `printf` statement in the first example program, but just to clarify here are two more example character strings:

```
"This is a character string"
"This contains \n a new line"
```

If a string constant is too long to fit on one line it cannot be continued on the next line. However, two adjacent character strings will automatically be joined to form a single string. These strings can be separated by any white space character, for example a newline. Thus in the following two statements the first is not valid, while the second is.

```
printf("This is an
        illegal string\n");

printf("This is a perfectly"
       " valid string\n");
```

The concatenation (joining) of adjacent character strings was not defined in the original C language. Many implementations did, however, support this before it was formalized by the ANSI standard.

Summary

- Integral constants are any sequence of digits which result in a number within the permitted range.

- Integral constants may have the suffix 'U' to indicate `unsigned`, and/or the suffix 'L' to indicate `long`.

- An integral constant beginning with a zero is interpreted as octal.

- A real constant is of type `double` if there is no suffix, of type `float` if the suffix is 'F', and of type `long double` if the suffix is 'L'.

- Character constants are represented either as an integral number within the permitted range, or as a single character within single quotes (e.g. `'A'`).

- Special characters are represented using an escape sequence where the first character is a backslash. For example, a backspace character would be represented by `'\b'`.

- A string constant is a sequence of characters (including, if required, backslash escape sequences) enclosed within double quotes (e.g. `"Hello\n"`).

- A string constant contains a terminating zero byte.

Exercises

1 If 33 is a constant of type `int` (or `char`) what are the type(s) of the following constants?
3.0
2L
'A'
3.5L
0x11
4.1F

2 What are the decimal values of the following constants?
11L
015
0x10

5

Variables

Aims and objectives

The aims of this chapter are to:

- define permitted variable names;
- show how variables are declared;
- distinguish between automatic and static variables;
- define variable scope;
- demonstrate how to preset variables.

Variable names

Variables are named entities which can hold data of a given type. Names can be of any length but the ANSI standard only requires the first 31 characters to be significant. To be honest, names this long can be more of a hindrance than a help. Variable names can contain any mixture of letters (upper and lower case characters are distinct), numbers and the underscore character (_). The first character in a name must not be numeric. Also the underscore character should not be used as the first character of a name, as this has special significance in standard library functions.

Certain names are reserved by the compiler and must not be used. These are shown in Fig. 5.1.

Fig. 5.1 Reserved names

auto	break	case	char
const	continue	default	do
double	else	enum	extern
float	for	goto	if
int	long	register	return
short	signed	sizeof	static
struct	switch	typedef	union
unsigned	void	volatile	while

Variable declaration

Variables hold data of a given type and before any variable can be used its type must be explicitly declared. This is done using a type declaration statement containing the type and variable name(s) as follows:

```
int count;
char initial, flag;
double pi;
```

Variables should all be declared at the start of each function, before any executable statements are encountered.

Automatic variables

There are two sorts of variable, automatic and static. Unless specified otherwise, variables declared within a function are automatic. This means that the variables are in existence only during the time that the function in which they are declared is being executed. Thus a variable is known only within the function in which it was declared. Another variable of the same name but in another function is totally independent. The value of an automatic variable is not maintained between calls to the function.

Static variables

On the other hand a static variable occupies a fixed location in memory which does not change as the program is executed. Therefore the value of a static variable is retained between function calls. Variables declared outside any function are static. Variables declared within a function can be specified as static by preceding the type declaration with the keyword `static` as in:

```
static int in;
static float var;
```

Initializing variables

When variables are declared they may be preset to an initial value. This is done as follows:

```
int count = 0;
float sum = 39.5F;
static double pi = 3.14159;
```

Each time a function is called, automatic variables are set to their preset value. Automatic variables which are not preset contain an undefined value. Using an undefined value in a calculation could produce erroneous results.

Static variables are preset only once, when the program is loaded. Static variables which are not preset are initialized to zero.

Global variables

Variables declared outside of any function are static. They are also global, which means that they can be accessed by any function defined subsequently. Global variables can also be accessed externally, that is, by functions compiled separately. This feature is not covered further in this text as it will not be required for any of the projects. Those who feel the need should refer to the `extern` keyword in their reference manuals.

The keyword `static` on global variables removes their global attribute. The variables can still be accessed by any subsequent function, but not externally. As splitting a program into several separate source files is not covered in this text, this usage of the `static` keyword can be ignored.

Variable location

It may help to remember the difference between static and automatic variables, and how they are initialized, if the location of the variables is explained.

All static variables, that is, global variables and variables declared as static, are stored in an area of memory called the heap. This is in a fixed location of memory, as far as the program is concerned, and preset variables which are in the heap are initialized when the program is loaded into memory, before it starts to execute.

On the other hand, automatic variables are located on the stack. This is usually a smaller area of memory which grows and shrinks as the program is run. When a function is called any automatic variables are allocated space on the stack and, if preset, then their values are set. If the same function is called a second time the automatic variables will usually not occupy the same position on the stack. Automatic variables which are not preset contain undefined values.

Variable scope

The portion of code within which a variable can be used is known as its scope. If a variable is declared within a function, then the scope of the variable is from its declaration, to the end of that function. The variable can only be referenced within that function. However, a variable declared outside of any function can be used in any subsequently defined function. To summarize scope, consider the section of program shown in Fig. 5.2.

Fig. 5.2   Sample program section

```
int num = 0;
main()                          /*   main function   */
{
   int i;
      .
      .
      .
}

char x;

func1()                         /*   function one   */
{
   int i = 1;
   static double pi = 3.14159;
      .
      .
      .
}

func2()                         /*   function two   */
{
   int i, num;
      .
      .
      .
}
```

Taking each variable in turn, the first (num) is global. It can be accessed by all functions except func2, which cannot use it because this function has a local variable of the same name. A local variable is used in preference to a global variable with the same name. The global variable, together with the functions, can be accessed by routines compiled separately, although system restrictions may limit their names to as few as six characters.

Each of the three functions contains a local variable i. These are independent and exist only while the function is executing. In func1 the variable i is preset to one every time the function is called. Also in this function the variable pi is preset. But, as this variable is static, it is only preset once, when the program is loaded. The variable is only accessible within func1, but its value is retained from one call to the next.

The character variable x is global but is known only by the function func1 and func2. The function main cannot address this variable.

Summary

- Variables are named entities which can contain data of a given type.

- Variable names consist of alpha-numeric characters (including the underscore) and should neither exceed thirty one characters in length nor start with a number. Using an underscore character at the start of a name has a special meaning and should be avoided. Certain names must not be used as they are reserved by the compiler.

- The area of a program within which a variable may be used is known as its scope. The scope of a variable declared within a function is that function. The scope of a variable declared outside of a function is from its declaration to the end of the file.

- Automatic variables (i.e. those declared within a function and without the static keyword) are in existence only while the function in which they are declared is being executed. On the other hand, static variables exist, and retain their value, whatever functions are called.

- Unless preset, automatic variables have undefined (i.e. arbitrary) initial values. Static variables that are not explicitly preset are initialized to zero.

Exercises

1 Which of the following variable names are invalid, and why?

```
maximum_result
2pi
bill.
summary
register
```

2 What are the initial values of all the variables in the following section of code?

```
int i;
five()
{
   int j;
   static int k;
}
```

3 If a variable defined within a function is of type static, is the value of that variable maintained between calls to that function?

4 Are global variables static?

6

Operators

Aims and objectives

The objectives of this chapter are to cover:
- the C operators;
- operand casting;
- operator precedence.

Introduction

Compared with other languages, C has a large number of operators, and this is often highlighted by proponents of other languages. This argument carries little weight, however, as many operators are combinations of others. These may be used as a shorthand if preferred. The basic operator types are: arithmetic, bitwise, logical, relational, assignment, and miscellaneous. Before looking at the operators in detail, there is a comment on the operands.

Casting

Where an operator requires two operands (as in a + b) then both operands should be of the same type (e.g. both int or both float, etc.). Where the operands are of different types then one operand should be converted to the same type as the other. This conversion is known as casting and is accomplished by preceding the operand to be changed by the required type enclosed within parentheses. For example:

```
float f = 3.8F;
int i;
i = 2 * (int) f;
```

The resulting value for i will be six as the fractional part is dropped. Note that the value of f is not changed, only the value used in the calculation. If operand casting is required, but not explicitly done, then the compiler will do the required conversion automatically. This automatic conversion will be done by raising the lower of the two types to that of the upper. Thus char will be promoted to int, int raised to float, and float raised to double. What happens to unsigned types is implementation dependent. However, there is no need to learn the rules as the compiler should issue a warning message saying operands are of a differing type. This can then be corrected by inserting the required cast.

·A word of warning. Consider what would happen in the previous example if the variable f were not cast. The expression would be evaluated using real numbers (making the result 7.6) which would then be converted to an integer with the value seven (the fractional part being dropped, not rounded up). With the cast the value is six, without the cast it is seven. Wherever possible it is worth avoiding any form of casting within an expression, that is, keep all operands of the same type, but when required do the casting explicitly and double check.

Compiler warning messages should be treated as error messages because, more often than not, unexpected results will occur when they are ignored. Beware of compilers which have different levels of warning message and often suppress these warnings. This is more of a hindrance than a help. When using Microsoft C compilers always select message level three, that is, print all error and warning messages. A good program will not generate any!

Arithmetic operators

The four standard operations of addition, subtraction, multiplication, and division are available using the symbols +, -, *, and /. The normal operator precedence applies, that is, multiplication and division are done before addition and subtraction, otherwise the symbols are evaluated from left to right. Parentheses can be used to alter the evaluation order. Thus:

```
3 * 2 + 3 * 3          evaluates to 15
3 * (2 + 3) * 3        evaluates to 45
```

There is also a modulus or remainder operator %, so that 5 % 2 evaluates to 1. The precedence of % is the same as * and /. Appendix B shows the operator precedence, or hierarchy, for all operators. Note that most operators are evaluated left to right but some are processed right to left.

It should be pointed out here that operator precedence is a useful concept to determine the correct evaluation order of an expression. However, the C definition uses a grammar, and does not use precedence. Thus our use of precedence order is a useful aide memoire, but not a rigorous definition.

Auto–increment and decrement operators

Frequently within a program there is a need to increment or decrement the value of a variable by one. The two operators for this are ++ for increment, -- for decrement. These operators have a higher precedence than * and /, hence will be done first. The ++ and -- operators can be positioned either side of the variable to which they belong, for example:

```
++i
i++
```

However, these two expressions are not equivalent. If the operator precedes the associated variable name, then the increment or decrement is done before the variable is used in the expression. If the operator comes after the variable name, then the increment or decrement is done after the value of the variable has been used.

Consider the two expressions (i++ * 3) and (++i * 3), where the initial value of i is two. After either expression has been evaluated the value of i will be three. However the first expression has an overall value of six, while the second produces the value nine. Thus if the initial value of i is 2 then:

```
i++ * 3        is   6
++i * 3        is   9
```

Bitwise operators

Bitwise operators are used to perform bit manipulations of integral data (i.e. chars and ints). They are available because C is a general purpose language and programmers may well wish to do such things. In this text bitwise operators will be used but not that frequently. Bitwise operators are most easily described using the ones and zeros of binary notation.

Fig. 6.1 Binary bit values

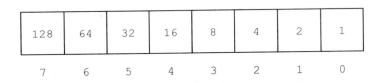

128	64	32	16	8	4	2	1
7	6	5	4	3	2	1	0

Consider a single byte containing eight bits each of which can be either a zero or a one. The bits are labelled from zero to seven starting at the right. A decimal value can be assigned to each bit position calculated as 2^n, where n is the bit position. This gives the decimal values as shown in Fig. 6.1. The value of a byte is the sum of the decimal values for which the bit is set to one. For example 00000101 is five.

AND (&)

The bitwise and operator (&) takes two operands. Where the corresponding bits of both operands are set to one the result is also set to one. Thus:

AND
_____

```
        00000101    (5)
    &
        00000110    (6)
    =
        00000100    (4)
```
_____

OR (|)

The bitwise or operator (|) has a one bit in each position of the result where either or both corresponding bits of the two operands are set. Thus:

OR
_____

```
        00000101    (5)
    |
        00000110    (6)
    =
        00000111    (7)
```
_____

Exclusive OR (^)

The exclusive or operator (^) sets each bit of the result where either, but not both, corresponding bits of the operands are set. Thus:

XOR
_____

```
        00000101    (5)
    ^
        00000110    (6)
    =
        00000011    (3)
```
_____

Shift (<< and >>)

There are two shift operators, << for left shift, and >> for right shift. Each shifts the first operand by the number of bits specified by the second. Vacated bit positions are set to zero. Thus:

Left shift

`<<`	`00000101`	`(5)`
	`2`	
`=`		
	`00010100`	`(20)`

Right shift

`>>`	`00000101`	`(5)`
	`2`	
`=`		
	`00000001`	`(1)`

Left shift is equivalent to multiplying by 2^n where n is the second operand, providing the number does not run off the end, that is, overflow the available space. The right shift is equivalent to dividing by 2^n.

Shifting negative numbers

Beware of right shifting negative numbers. A signed number usually uses the most significant bit (bit seven in our example) as a sign flag. If this bit is set then the number is deemed to be negative. When shifting right this bit can either be kept as one or set to zero. This is not defined in the standard. You have been warned. Another thing not defined by the standard is what happens if the number of bits to shift is either negative or larger than the number of bits in the variable. Both cases should be avoided.

Ones complement (~)

Finally a single unary bitwise operator, which is the ones complement. This operator (~) changes all ones to zero and all zeros to one. Thus:

Complement

`~`	`00000101`	`(5)`
	`11111010`	`(250)`

Note that the above result of 250 assumes an unsigned value. If the value were signed then the result would be −6 if twos complement format were used, or −5 for ones complement form.

Bitwise operator precedence

As with arithmetic operators there is an order of precedence. Except for the ~ all bitwise operators have a precedence lower than the arithmetic operators. The precedence is then, shift operators, and, xor, or in descending order. Appendix B gives the full precedence order.

Relational operators

Relational operators compare two operands and return a true or false value, true being a value of one, false being a value of zero. They are usually used in conditional tests but there is no restriction to this effect. The permitted operators are listed in Table 6.1.

Table 6.1   Relational operators

Operator	Meaning
<	Less than
<=	Less than or equal to
==	Equal to
!=	Not equal to
>=	Greater than or equal to
>	Greater than

Logical operators

There are three logical operators. These are generally used to either combine or negate the results of relational expressions.

To test if two or more relational expressions are all true, the logical and (&&) operator can be used. For example:

```
(a < b) && (b == c)
```

Only if both relations are true will the overall result be true. The parentheses are optional as the precedence of && is lower than either < or ==, but it does make the expression easier to read. To test if one or more relational expressions out of a group is true use the logical or (| |), for example:

```
(a < b) || (b++ > c)
```

If either relationship is true then the result is true. Note that C works from left to right and will stop evaluating expressions as soon as the result is known. Thus in the above example, if a is less than b then the second relational test will not be done, and b will not be auto–incremented.

There is also a single unary logical operator ! which converts a non-zero value (i.e. true) to zero (i.e. false) and vice versa. For example:

    !(a < b) is equivalent to (a >= b)

Assignment operators

The first, and most obvious, operator in this section is the assignment operator (=). This operator evaluates from right to left so that the right–hand expression is evaluated and this value is then assigned to the operand on the left. Normally the left operand is a simple variable (known in error messages as an 'lvalue'), but can be any expression which describes an object whose value can be set.

This is a modification introduced by the ANSI standard. In earlier versions of C the left operand had to be a simple variable. More complex structures, which could take several values simultaneously, were not permitted.

As the assignment operator acts from right to left there can be multiple assignments in a single statement. Thus the following is perfectly valid:

    a = b = c = i + 2;

Automatic casting

The types (i.e. int, char, float, etc.) of both operands should be the same. If they are not of the same type then the compiler may issue a warning message, but will add the necessary code to convert the right–hand operand to the type of the left. This may lose resolution. For example:

    int i;
    i = 2.5;

This is perfectly valid and will result in the variable i being assigned the value two. The casting from type double into int being done by the compiler. Hopefully a warning message would have been printed and the possible error spotted.

Combined assignment operators

A common construct in programs is i = i + 2;, or something similar, where an assignment expression has the same variable in both operands. As a form of shorthand any arithmetic or logical operator using two operands can be combined with the assignment operator. Thus the previous expression can be rewritten as i += 2;. Whilst in the earlier days this may have helped the compiler produce more efficient code, this should not be true now. The code produced by a good compiler should be identical for either source form. These combined operators just form a convenient shorthand.

As a word of warning, compilers sometimes get confused if there is not a space either side of a combined operator. This results in a message about operator ambiguity. A general rule is space expressions out; they are also often more readable.

Miscellaneous

The operators in this section do not fit into the preceding categories but are nevertheless useful.

Conditional operator

The conditional operator (?:) is rather odd. This operator evaluates its left operand and, depending on whether the result is true or false, sets the overall value to one of the two right–hand operands. These operands are separated by a colon. Thus if the left operand is true (non-zero) then the first of the two right–hand operands is the value, otherwise the second value is taken. For example:

```
c = (a < b) ? 5 : -5;
```

If a is less than b then the variable c is set to the value 5, otherwise it is set to the value −5. The if statement, which will be described later, provides the same functionality but in a more readable form; however, the conditional operator does provide a useful shorthand on occasions.

Comma operator

Next the comma operator (,) which is used as a delimiter when more than one expression is to be combined into a single statement. Generally it is not good form as the resulting code tends to be dense and cluttered, but it exists and now you know about it. As an example the following is perfectly valid:

```
a = 3, b = 6;
```

The assignments could have been equally well written as:

```
a = 3; b = 6;
```

The difference is that the first example is a single statement containing two assignment expressions, while the second example contains two statements, each of which is an assignment expression. There are very few places where a comma operator is useful. Generally it adds unnecessary confusion.

`sizeof` operator

This operator is used to determine the size, in bytes, of a particular variable, or variable type. For example, to determine the number of bytes used to store the variable x one would use:

```
int i;
i = sizeof x;
```

Also a variable type can be specified rather than a variable name. To do this, the required type must be enclosed within parentheses. This can be thought of as a cast. For example:

```
int i;
i = sizeof(float);
```

One minor point needs clarifying here. The `sizeof` operator 'creates' a numeric value from the supplied operand. The type of this value is implementation dependent, and is known as `size_t`. It is always an unsigned integral type, often `unsigned int` or `unsigned long`. In general, the result of a `sizeof` operator is a small number which will fit into any integral type.

Summary

- The C operators can be grouped into the following categories: arithmetic, bitwise, logical, relational, assignment, and miscellaneous.

- Where an operator requires two operands, these should both be of the same type. If they have different types then one, or both, should be explicitly cast so that the types match.

- Casting is used to change the type of an expression. It is achieved by putting the required type, within parentheses, immediately before the expression.

- Where casting is required, but not explicitly stated, the compiler will automatically insert a cast. It does this by promoting the lower type, thus avoiding loss of resolution.

Exercises

1   For each expression below determine the resulting values, given that—for each expression—the initial values of i, j, and k (each of type int) are 1, 3, and 5 respectively, and variable f (of type float) is 3.5.

```
i++ * j

(float) k * f

k * (int) f

i<j ? j : k

i & k << j

i++ + j++ + ++k

sizeof i
```

7

Formatted output

Aims and objectives

The aims of this chapter are to:
- introduce the standard library concept;
- show how to output messages to the screen;
- describe how to print the values of variables.

Introduction

Although the C language does not have any input or output directives, there is a large suite of standard library functions available to do just about everything one ever wanted, plus a bit more besides. This chapter deals only with one function `printf` as this is the cornerstone of all program development, it being one of the best tools for debugging!

printf

The function `printf` writes output to standard output. This is the default output device and is usually the screen. Some computers still have only line printer output, but their days are numbered. The first example program used this function to write a message, but now a full description is required. The general form of the `printf` statement is as follows:

```
printf("format string" [,var1] [,var2] [,...]);
```

The items enclosed within square brackets are optional. The square brackets are not included in the function call.

Format string

The first item is a format string and specifies how the output is to be written. It can be a string variable, which has yet to be defined, or a string constant as in our example program. The function will print only what it is given in the format string, that is, no added end of line characters. The compiler processes the backslash character (\) as described earlier under character constants thus enabling non-printing characters to be output. A quick glance at the very first program (Fig. 1.1) shows this usage to output a newline.

printf itself processes one character in a special manner. The percent sign (%) is used to indicate that the value of a variable is to be inserted at this position. Use two adjacent percent signs if one is to be printed.

Conversion characters

The percent character indicates that the next variable from the list of parameters following the format string is to be printed at this point. The key letter indicates what form this variable is, that is, int, float, and so on. The permitted key letters are shown in Table 7.1.

Table 7.1   Format key letters

Letter	Meaning
%c	Single character
%d	Integer as decimal
%e	float or double in form m.nnnE+xx
%f	float or double in form m.nnn
%g	Smaller of %e and %f
%o	Integer as octal
%p	Pointer notation (see later)
%s	Character string
%u	Unsigned decimal integer
%x	Integer as hexadecimal
%%	Percent sign

Field width

The key letter may be prefixed by a numeric value—as in %5d—which indicates the required number of columns the number is to occupy. The number will be right justified in these columns. If the number is too large to fit in the specified number of columns, the number is printed anyway, using as many columns as required. There is nothing worse than a program which takes several hours to run, failing to produce any output because too small a field width was specified. If no width is specified only those required will be used, no spaces or filling characters will be inserted. Thus a typical use for the printf function would be:

```
printf("The  number  is  %d\n",  num);
```

Leading zeros

If the number after the percent sign starts with a zero, then the number is not octal, as one might expect, but indicates leading zero characters are to be printed and not suppressed as would be the norm. Thus the number seven printed using %03d would be printed as 007.

Resolution

For floating–point values the number of decimal places can be specified by giving the required count following a decimal point. Thus %8.3f would output a floating–point number using eight columns and having three digits after the decimal point.

Long variables

If either a long int or a long double variable is to be output then the key letter should be prefixed by the letter l. For example %d would become %ld. The result of omitting this additional character is not defined, but usually a truncated value is output.

Caution

printf gets confused if the number of variables does not match the number of format specifiers given. Too many parameters do not matter, but too few cause chaos! Another thing that can throw printf completely is to print values with inappropriate format descriptors, for example printing an integer using %f .

Return value

Like most standard functions printf returns a value. In this case the value is of type int and is a count of the number of characters written. If the count is negative then an error occurred. printf so rarely fails that the return code is usually ignored.

Summary

- The printf function writes to the default output device—usually the screen.

- The first parameter specifies how the output is to look and is printed verbatim, except for the backslash (\) and percent (%) characters.

- The backslash character is processed, by the compiler, in the same way as for character constants. This enables non–printing characters (e.g. tabs and newlines) to be output. Note that it is not `printf` that processes these codes, it is the compiler.

- The percent sign indicates to `printf` that the value of a variable is to be inserted in the output. The character(s) after the percent sign indicate the type of this variable and in what form it is to be printed.

- Use two adjacent percent signs if one is to be printed.

Exercises

1   Try modifying the original program by adding a few `printf` statements of your own. Try printing a few variables which have been preset and also see what happens when errors are made. Use incorrect format descriptors and/or leave out variables when required. Becoming familiar with the type of diagnostic message issued by the compiler will help with real programs later.

2   Write a program that uses the `printf` statement, together with the `sizeof` operator, to print the size (in bytes) of the various variable types available. The following would be a good starting point.

```
printf("char is %d bytes\n", (int) sizeof(char));
```

Note the use of a cast before the `sizeof` operator. As mentioned earlier, the `sizeof` operator returns a value of type `size_t`. The definition of this type depends on the particular compiler being used, and can be any unsigned integral type. Specifying the cast enables the format descriptor `%d` to be used. Without the cast it might need to be `%u` or `%lu`, depending on how `size_t` is defined.

8

Structured programming

No one knows what the words 'structured programming' mean, exactly. Yet we find those arguing strongly in its favour and those arguing strongly against it. How does one argue for or against something when one is not even sure what it is?

P. J. Denning (1974)

Aims and objectives

The intention of this chapter is to:

- show that C is a suitable language for structured programming.

Overview

For the purposes of this text the term 'structured programming' will be taken as a method of producing a concise and clear solution to a problem. While theorists may throw up their hands in horror, this appears to be a reasonable precis of the myriad of texts on the subject.

The two general approaches to structured programming—namely 'top down' and 'bottom up'—are discussed in the project introduction (Chapter 30). For the moment we will concentrate on the language rather than its usage.

Language requirements

Proponents of structured programming recommend a certain group of control constructs. These are:

```
if (some condition is true)
    do some statements
otherwise
    do some other statements

while (some condition is true)
    do some statements
```

```
repeat
   some statements
while (some condition is true)
```

C supports all these forms and adds two more for good measure. These five control statement constructs are detailed in the following chapters.

It should be remembered that while these constructs allow good programming practice, they do not enforce it. It is the job of the programmer, not the language, to make a good program.

The infamous goto

A common feature among proponents of structured programming is their hatred of any form of jump or goto statement. Indeed, one such author has suggested that these statements should be withdrawn from all high–level languages. While C does not go quite this far, the author recommends avoiding the use of goto whenever possible. To this end discussion of the statement, and the ANSI longjmp function call, is deliberately omitted. None of the projects requires these statements. If desperate to use them reconsider, and then consult your reference manuals.

Summary

- C has all the required constructs for structured programming.

9

---

if

Aims and objectives

The aims of this chapter are to:

- describe the form of the if statement;
- describe the optional else keyword;
- show the usage of compound statements.

Syntax

The if statement is one of only two conditional statements available in C. An expression is evaluated and the following statement executed if the expression returns a TRUE (i.e. non-zero) value. For example:

```
if (i < 5)
    printf("The value of i is less than 5\n");
```

This demonstrates the most common usage where the test expression involves a relational operator. There is no reason why a simple variable or expression cannot be used as in:

```
if (first)
    printf("The value of first is not zero\n");
```

As stated earlier, any single statement may be replaced by a compound statement, so if more than one statement is required under the control of an if then use the following form:

```
if (i < 0) {
    printf("i was negative, sign changed\n");
    i = -i;
}
```

Often the need arises to execute one statement if some condition is met, otherwise execute a different statement or statements. The else keyword allows this as follows:

```
if (i < 0)
  printf("i is negative\n");
else
  printf("i is positive (or zero)\n");
```

Remember that either or both of the single statements may be replaced by a compound statement if required.

Layout

Note how indentation has been used to highlight which statements the if controls. This is only for readability, as the compiler ignores indentation. Everything could be written on one line but it then becomes far less readable. It is very important to make things readable.

There is no fixed rule on exactly how indentation is done, though there are a number of (different) guidelines. This text uses two spaces for each level of indentation. Where braces are used, the opening brace is at the end of the line containing the if keyword. The closing brace is on a line of its own, and is in the same column as the letter i of the if keyword.

Real tests

Avoid tests for equality using non-integer values. This is common in all languages and is a result of possible rounding errors causing unexpected results. For example 10.0 / 2.0 might give 4.9999999999 rather than 5.0 thus the test (10.0/2.0 == 5.0) could fail.

Nested ifs

Be careful when putting one if statement within the bounds of another, when one of the ifs has a matching else. As an example consider the following and determine the possible actions.

```
if (i < 0)
  if (j < 0)
    printf("Both i and j are negative\n");
else
  printf("i is not negative, j not checked\n");
```

An if together with its associated statement, plus any else and its associated statement, are taken as a single statement, but in the above example to which if does the else belong? Do not forget that the compiler ignores the indentation. Try it and see. You should find that with both variables as zero or positive there is no output. In other words the else has matched the nearest if, not as is implied by the indentation.

To have the desired effect, and as a by–product to make it easier to read, include the second `if` within a compound statement as follows:

```
if (i < 0) {
   if (j < 0)
      printf("Both i and j are negative\n");
}
else
   printf("i is not negative, j not checked\n");
```

Modify your test program and verify it now works as expected. Remember indentation is for you—it is ignored by the compiler.

Summary

- The general form of an `if` statement is:

```
if (condition)
   statement
else
   statement
```

- The `else` keyword and the associated statement are optional.

- When more than one statement is to be controlled using an `if` statement then use a compound form as in:

```
if (condition) {
   statement1
   statement2
   . . .
}
```

- Try to avoid testing for equality when using real numbers.

- Unless indicated otherwise using braces, an `else` is associated with the nearest `if`.

Exercises

1 Write a short program which uses an `if` test to print one of two messages depending on the value of a preset integer variable. Try changing the value of the variable and verify the test is done correctly.

2 Write an `if` statement to set the variable `i` to the smaller of the two variables `j` and `k`.

3   Repeat the previous exercise but this time use the conditional operator ? :
(described in Chapter 6) rather than an if statement. Is the code more
readable? A good compiler will generate identical code so the choice is
purely personal.

10

---

switch

Aims and objectives

The aims of this chapter are to:

- describe the form of the switch statement;
- introduce the break keyword.

Syntax

The switch statement is used in preference to a sequence of if statements where one of a number of different statement groups has to be executed depending on the value of an expression. This is probably best demonstrated by using an example, so one is shown in Fig. 10.1.

Fig. 10.1 Example switch statement

```
switch (i) {
   case 1:
            printf("i is 1\n");
            break;
   case 2:
   case 3:
            printf("i is either 2 or 3\n");
            break;
   case 4:
            printf("i is larger than 3\n");
   case 5:
            printf("i is 5 (or 4)\n");
            break;
   default:
            printf("Unknown value of i\n");
   }
```

Following the keyword switch there is an expression enclosed within parentheses. This expression, which must produce an integral value (e.g. char, int, short, or long), is used for the subsequent cases. The body of the statement is enclosed within braces and includes one or more `case` keywords, each of which is followed by an integral constant and a colon. If the value of the `switch` expression matches the value of a `case` then the statements following the relevant `case` are executed. All subsequent statements will be executed until either a `break`, or the end of the `switch`, is reached.

Note the default case: this is a catch-all and will be processed if none of the previous cases matches. If used, the definition of the default case is usually placed last. If none of the case values matches the expression, and there is no default case, then the `switch` statement is effectively ignored.

Go through the above example with the values of i in the range one to six. Note particularly the effect with a value of four. The omission of the `break` statement is a common and useful feature.

Where a `break` statement between cases has been deliberately omitted, it is probably worth commenting that it is intentional, as, when later debugging the program, this intention may be forgotten and a `break` inserted in error.

In the original definition of C the switch expression had to be of type `int`. The ANSI standard now allows any integral type. Beware, some compilers still allow only expressions of type `int` even though they claim ANSI compatibility.

Summary

- The general form of the `switch` statement is:

```
switch (expression) {
   case const₁:
                statement
                statement
                . . .
                break;
   case const₂:
                statement
                statement
                . . .
                break;
   default:
                statement
                statement
}
```

- The `switch` statement is used in preference to a sequence of `if`s.

- The expression value, and the constants associated with each `case` keyword, must be integral.

- The `default` keyword begins a sequence of one or more statements which will be done if no cases match.

- If used, the default case is normally placed at the end of the `switch` statement.

Exercises

1 Write a `switch` statement to write out whether a given integer number is odd or even. Assume that the numbers will be in the range zero to nine and use only two `printf` statements.

2 Write another `switch` statement where the switch expression is the reply to a question requiring a yes or no answer. Do not worry about how the input is done, just assume that the first character of the reply is available in a variable of type `int`. Remember each case expression can be a character constant. Allow for both upper case and lower case replies and also check for an invalid response by using the `default` keyword.

11

while

Aims

The aims of this chapter are to:
- describe the syntax of the `while` statement;
- introduce the `continue` keyword;
- define a second meaning for the `break` keyword.

Syntax

The `while` statement is used to repeat execution of a statement or statements while some logical expression has a true (i.e. non-zero) value. The form of the `while` statement is:

```
while (condition)
   statement
```

As an example, consider the following program segment that prints the numbers from one to ten inclusive:

```
int i = 1;
while (i <= 10)
   printf("%2d\n", i++);
```

As mentioned earlier, C is a free format language so the while statement could have been put on one line, as in:

```
while (i <= 10) printf("%2d\n", i++);
```

However, the first form is probably clearer, especially when it is remembered that a single statement can be replaced by a compound statement. This would be used when more than one statement was required within the loop, for example:

```
while (condition) {
   statement1
   statement2
      . . .
}
```

Again the format is not compulsory, but the form shown, with indentation highlighting those statements within the body of the loop, is clear and will be used throughout this text.

Now we know what a `while` statement looks like, let's see how it works. First the condition expression is evaluated and, if the result is true (i.e. non-zero) then the statement forming the body of the `while` is executed. The condition expression is then re-calculated and the loop repeated until a false (i.e. zero) value is returned by the conditional expression. Note that if the condition is never true then the body of the loop will never be executed.

Null loops

A common mistake when using a `while` loop with a single statement is to add an extra semicolon after the condition as in:

```
while (condition);
   statement
```

Although this may look correct the additional semicolon creates a null loop, that is, one containing no statement. The statement following is always executed as it is not under the control of the `while`. The compiler will not create an error as the syntax is correct.

A null loop is sometimes used intentionally. When this is the case, the author recommends putting the semicolon on a line of its own as in:

```
while (condition)
   ;
```

Example `while` usage

Now for a real example that prints the squares of the numbers from one to ten inclusive. Enter the program shown in Fig. 11.1 and verify that it works.

Fig. 11.1   Example `while` loop

```c
#include <stdio.h>

main()
{
   int i = 1;
   while (i <= 10) {
      printf("%2d - %3d\n", i, i * i);
      i++;
   }
}
```

Counting from zero

In natural languages counting usually starts at one, a zero value being ignored. But to computers zero is a perfectly useful number so counting round a loop ten times would usually be done using:

```
int i = 0;
while (i < 10) {
   .
   .
   .
   i++;
}
```

Terminating early

If the conditional expression is a non-zero constant then an infinite loop will result. This is not as silly as it may seem because there are a number of ways to get out of a loop. Ignoring the infamous goto and deferring discussion of the return statement until the section on functions, still leaves break and continue.

break

The break statement causes execution of the innermost loop to be terminated. As soon as this statement is reached the remainder of the statements within the loop are skipped, and execution continues at the first statement after the loop.

Beware: the break statement has two distinct uses. It can end a case in a switch statement and it can terminate a loop, but it cannot do both at once. If a switch is contained within a loop then any break statements within the switch do not terminate the loop.

continue

The continue statement is similar to the break statement in that it causes any remaining statements in the loop to be skipped. However, the loop condition is then re-evaluated and, if true, the loop is repeated. Thus the continue statement causes a skip to the end of the loop, then a possible repeat, while a break skips to the end and terminates the loop.

Summary

- The general form of the `while` statement is:

  ```
  while  (condition)
     statement
  ```

- If the condition is never true then the statement(s) are not executed.

- The `break` statement terminates (i.e. jumps out of) a loop.

- The `continue` statement skips any remaining statements in the current loop iteration and then re-evaluates the condition. If this is still true then the loop is repeated.

Exercises

1   Modify the squares program shown in Fig. 11.1 so that the first line contains the square of ten and the last line contains the square of one. Try modifying the `printf` statement to produce output more to your liking.

2   Again, modify the original program this time to print the squares of the numbers from 1.0 to 2.0 in steps of 0.1. There is no reason why the operands used in the conditional expression should not be real numbers, but avoid tests for equality as rounding errors may cause unexpected results.

3   Finally if 10 000 pounds is invested in a savings account giving 6.5% interest, compounded annually, what will be the value of this account at the end of each year for the next twenty five years? Write a program to print the annual sums assuming that all interest is left in the account. As an un-needed hint, a `while` loop would form a suitable solution, and two variables are required, one for the loop count and one for the current invested sum. For those not familiar with degree level mathematics, the formula for calculating the new account value at the end of each year is given by:

new value = initial value + (initial value * 6.5) / 100.0

or (in C terminology)

```
sum += sum * 0.065F;            (Assuming type float)
```

12

for

Aims and objectives

The aim of this chapter is to:

- describe the `for` statement;
- demonstrate the perils if non-integral loop counters.

Syntax

If one considers how a `while` loop is used in real life, one often discovers three recurrent features. Before the body of the loop there is some initialization. Then there is a test to determine if the body of the loop is to be executed. Finally there is often an increment, decrement or other simple expression executed at the end of each loop. For example, consider the following `while` statement:

```
i = 0;
while (i < 10) {
    .
    .
    .
    i++;
}
```

The three parts are: `i = 0;` initialization, `i < 10` condition, and `i++;` increment.

The `for` loop enables these three parts to be combined into one statement as follows:

```
for (i=0; i < 10; i++) {
    .
    .
    .
}
```

As always either a single statement or a compound statement may be used, the above example uses a compound one. Note how each part of the for statement is delimited using a semicolon. Normally this character is used to indicate an end of statement; however, its use here is not inconsistent.

Any one or more of the fields can be a list of expressions separated by the comma operator. Thus the following is perfectly valid:

```
for (i=0,j=10; i < j; i++,j--)
    printf("The value of i is %d\n", i);
```

Consider the above and confirm that five values will be printed, zero to four.

Terminating a for loop

Note that the three fields in a for statement are independent. They are also all optional. If the condition is omitted then it is assumed to be true. The following is a well–used infinite loop:

```
for (;;) {
    .
    .
    .
}
```

To terminate a for loop early, that is, while the loop condition is still true, the break statement can be used as described under while.

Similarly the continue statement can be used to skip the remainder of the body of the loop and cause the conditional expression to be re-evaluated. Note that any increment or decrement from the loop header will still be done after a continue statement is executed. For example, the following code will print the numbers from zero to nine excluding five:

```
for (i=0; i < 10; i++) {
    if (i == 5)
        continue;
    printf("%d\n", i);
}
```

When to use a while loop and when to use a for is really a matter of taste. Usually when counting is involved a for loop is simpler, otherwise generally a while is used.

Real expressions

The expression fields are not restricted to integral values. It is perfectly valid to do the following:

```
double x;
for (x = 0.0; x <= 10.0; x += 0.2)
   printf("%4d  %f\n", (int)x, x);
```

Consider the previous example and predict the outcome. Then enter the additional lines required to make it a complete program and verify your prediction.

What may have happened—and this depends on the compiler and computer used—was that the integer value printed was not the same as the floating–point value. For example the last few lines of the output might have been:

```
9     9.4000
9     9.6000
9     9.8000
9    10.0000
```

This was caused by rounding errors. The number 0.2 (like many other real numbers) has no exact representation within a binary computer. Thus the nearest value is used, and this might well be 0.199999. This, as mentioned earlier, is why one should always avoid testing real numbers for equality.

However, within a loop these errors will accumulate. Not only is the value 0.2 not stored exactly, but also the result of adding this value to another will introduce a rounding error.

If the maximum accuracy is required in a loop construct it is better to use an integral loop counter and convert this value each time round the loop. This means that no cumulative error is introduced. For example, the previous program would be re-written as:

```
int i;
double x;
for (i = 0; i <= 50; i++) {
   x = (double) i * 0.2;
   printf("%4d %f\n", (int) x, x);
}
```

Summary

- The general form of the `for` statement is:

  ```
  for (initial; condition; expression)
     statement
  ```

- The initial expression is evaluated once, before the loop.

- The statement or statements forming the body of the loop are executed while the conditional expression returns a true (non-zero) value.

- The last expression is evaluated each time round the loop, after the statement(s) forming the body of the loop have been processed. This expression is often an increment or decrement.

- The `break` and `continue` keywords can be used with the same meaning as for the `while` loop.

- Real expressions are perfectly valid as the loop counter but beware that rounding errors will accumulate. If accuracy is important then use an integral counter and scale it to the required value each time round the loop.

Exercises

1   Write a program to print the squares of the numbers from 1.0 to 2.0 in steps of 0.1. Is this solution any clearer than the corresponding exercise in the preceding chapter?

2   Write a program that uses a `for` loop print the sum of the integer numbers from one to ten inclusive.

3   Modify the above solution to print two sums; one the sum of all the even numbers, and one the sum of all the odd numbers. When using integers the expression `i % 2` will return zero for an even number, and one for an odd number.

13

do

Aims and objectives

The aim of this chapter is to:

* describe the syntax of the do statement;
* suggest when to use the different loop constructs.

Syntax

The do loop (sometimes referred to as a do while loop) is the last of the three loop constructs. It is very similar to the while statement but has one major difference. Both the while and for loops test their condition and only if the condition is true is the body of the loop executed. If the condition is never true, then the body is never executed.

By distinction the do loop tests the condition after the body of the loop, hence it is always executed at least once. Therefore the condition indicates if the loop is to be repeated, rather than executed. The syntax of the do loop is:

```
do
   statement
while (condition);
```

Or, replacing the single statement by a compound one:

```
do {
   .
   .
   .
} while (condition);
```

Thus the body of the loop is executed then the condition is tested. If true, the loop is repeated. As an example, consider the earlier problem of printing the numbers from one to ten together with their squares. This could be accomplished using a do loop as follows:

```
i = 1;
do {
    printf("%2d - %3d\n", i, i * i);
} while (i++ < 10);
```

A word of warning. The last part of a do statement looks very similar to the first part of a while. The semicolon terminating a do statement could easily be forgotten. Conversely, as pointed out earlier, a semicolon after the while statement's condition would create a null loop, usually but not always an error.

The break and continue statements can be used within a do statement just as they can from within either a while or a for statement.

When to use a do loop

Now that there are three types of loop, what rules are there for determining which to use in a given situation? Generally all three are interchangeable but remember that a do statement always executes its body at least once.

The while loop is generally used when waiting for some condition to become true, for example reading data until there is none left to read.

A for loop is generally used when counting, for example to repeat a sequence of statements a given number of times.

The do statement is the least often used of the three loops. It is usually only found when the fact that the condition is tested after the body of the loop, is required. In other words some action is done, the result of which determines whether the action should be repeated. Waiting for a given character to be typed could be one such example.

Summary

- The general form of the do loop is:

  ```
  do
      statement
  while (condition);
  ```

- The break and continue keywords can be used as described for the while statement.

- The body of a do loop is always processed once.

- The condition expression returning a true (i.e. non-zero) value indicates the loop is to be repeated.

Exercises

1 Write a program that uses a do loop to print the sum of the numbers from one to ten inclusive.

2 If 10 000 pounds is invested in a savings account giving an annual interest rate of 6.5%, write a program to determine after what time the original investment would be doubled.

14

Functions (part one)

Aims and objectives

The aim of this chapter is to:
- describe why functions are used;
- define how functions are used;
- show how a value is returned from a function;
- show how values are passed into a function;
- introduce function prototypes.

Functions

As mentioned earlier, all programs are made from functions. These functions are either supplied as part of the standard library (e.g. `printf`) or are written by the user.

Why use functions?

The use of functions to split a large program into smaller, more manageable parts cannot be over–emphasized. These functions should perform a specific action and do it well. For example, a sorting program may have three functions, one for doing the input, one to do the sorting, and one for printing the results. Do not assume that functions should only be limited to sections of code which are to be repeated. Splitting a program into smaller units, even if each is only used once, can be a significant help in designing and debugging programs. To be of much use values have to be passed into and out of these functions, so let's see how this is achieved.

Return value

Most functions return a value, thus just as variables have type, so do functions. Unless specified otherwise the return value from a function is of type `int`. This default is a throwback to the original definition of C and has been retained for compatibility. As we shall see in a moment the return types for every function should be explicitly stated so this default should never be used. In order to return a value from a function the `return` statement is used. This has the form:

```
return value;
```

`value` can be a simple constant or any valid expression, providing the resulting number has the same type as that of the function. For example, to define a function of type `double`, which returns the value of the mathematical constant π, one could use:

```
double pi()
{
    return 3.141592653589;
}
```

Function usage

This is a perfectly valid function, possibly not the best way to define π, but it works. However, a problem can be seen for the compiler if one considers the program shown in Fig. 14.1 which uses this function.

Fig. 14.1   Simple function call

```
/*  Program to demonstrate function use   */

#include <stdio.h>

main()
{
    double p;
    p = pi();
    printf("The value of pi is %f\n", p);
}

double pi()
{
    return 3.141592653589;
}
```

Consider the above program from the point of view of the compiler. First it processes the function `main`. Here it knows variable p is of type `double` but at this stage has no knowledge of the function `pi`. When it reaches the statement 'p = pi();' it assumes `pi` returns an `int` value (the default) and will insert the necessary cast to convert the assumed integer value returned by the function `pi` into type `double` as required for the variable p.

Subsequently the compiler will process the function `pi`. It will discover the function is of type `double` which conflicts with its earlier assumption of `int`. Realizing the error it will try to shift the blame to the programmer by declaring an error in the definition of the function `pi`.

Function prototype

To get round this problem the ANSI standard introduced the concept of prototyping. A function prototype is a statement which is not executed but is used solely to inform the compiler about the types of variables passed into and out of a function. For example the function `pi` would have the following prototype:

```
double pi(void);
```

This statement, which would normally be placed outside of any function near the beginning of the program, informs the compiler that the function `pi` has no parameters (the keyword `void` means nothing), and returns a value of type `double`. Thus if this single statement was inserted in the previous example then all would be well. Try the program as shown in Fig. 14.2 to make sure that the usage and definition of the function `pi` does not generate any errors. Note that there will still be one or two warning messages about `main`, but for the moment these can be ignored. The prototype for the `main` function will be covered in detail in Chapter 22.

Fig. 14.2 Function prototype example

```
/*  Program to show function prototype use  */

#include <stdio.h>

double pi(void);

main()
{
   double p;
   p = pi();
   printf("The value of pi is %f\n", p);
}


double pi(void)
{
   return 3.141592653589;
}
```

Parameters

Suppose we now wished to extend the function `pi` so that it returned an integer multiple of the value π (e.g. 2π or 4π) where the multiple was specified as a parameter. The function prototype would need extending to indicate an `int` value was being passed into the function. This would be done as follows:

```
double pi(int);
```

The function would require modification to the following:

```
double pi(int n)
{
   return  3.141592653589 * (double) n;
}
```

The variable n is a function parameter. It can be thought of as a variable local to the function which has its initial value preset to a value specified when the function is called. To use this function one would need to modify the program shown in Fig. 14.2 to include the following:

```
p = pi(2);
printf("The value of two pi is %f\n", p);
```

Arguments and parameters

Now a bit of jargon. The value of two passed to the function `pi` in the above example is called an argument. The variable n within the function `pi`, which is passed the value of two, is known as a parameter. The value passed is an argument while the receiving variable is a parameter. Some texts use the term formal argument instead of parameter and actual argument instead of argument.

Pre–ANSI syntax

A word of warning. The ANSI standard introduced a number of changes in the area of functions and parameters. While this text always demonstrates the new definition, a knowledge of the old is still required to read existing programs. Firstly there were no function prototypes. Hence the types of parameters passed to functions could not be checked. The function return type had to be specified using a normal type declaration statement. A pair of parentheses after the name indicating it referred to a function rather than a simple variable.

Also the types of any function parameters had to be specified between the function header and the function body. Thus in the old style the previous example program would be as shown in Fig. 14.3.

Fig. 14.3   Pre–ANSI function usage

```
/*   Pre-ANSI function use   */

#include <stdio.h>

double pi();

main()
{
   double p;
   p = pi(2);
   printf("The value of two pi is %f\n", p);
}
double pi(n)
int n;
{
   return 3.141592653589 * (double) n;
}
```

Note how similar the type definition in the pre–ANSI program is compared with the corresponding function prototype in the ANSI example. The only difference is that the function prototype allows the number and types of any parameters to be checked, not only the return type.

Mixing ANSI and pre-ANSI functions

Beware of mixing ANSI and pre–ANSI format functions in the same program. There are small, but significant, differences in how arguments are passed and mixing them can lead to strange errors.

With pre-ANSI functions, arguments of type char and float were automatically cast to int and double respectively. For ANSI functions where a prototype defines an argument to be char or float then these are passed as defined and not cast. While this difference rarely matters, there are occasions where obscure program behaviour results. It is safer to stay with one form only, preferably ANSI.

Pass by value

Finally a bit more detail on parameter passing. Consider the function shown in Fig. 14.4 which raises an integer value to an integer power (e.g. 2^8).

Fig. 14.4   Parameter modification example

```
int power(int x, int n)
{
    int p;
    for (p=1; n>0; n--)
       p = p * x;
    return p;
}
```

Notice that the value of the second parameter (n) is altered within the function. This is perfectly valid but does not change the value in the calling function. This is because, as stated earlier, each parameter can be thought of as a local variable whose initial value is specified from the calling function. Any changes to the value within the function do not change the value in the calling function.

This parameter passing method is called 'pass by value' and tends to isolate functions better. There are arguments for and against this technique, but pass by value tends to produce code where functions are more independent, and less likely to produce unexpected side effects.

Pass by address

The inverse to this is 'pass by address' which is used, for example, by the Fortran language. Here if a function (or subroutine) modifies a parameter then the value in the calling function is also changed. When required this functionality is available in C as will be described later.

The main function

All the example programs tried so far will probably have generated either one or two warning messages when compiled. One may say that main has no prototype, and the other that main has no return value. The main function will be covered in detail later but for now, to prevent these warning messages all subsequent programs should end with a return statement (returning a zero value) and also include a prototype of the form:

```
int main(void);
```

Summary

- Functions are used to divide a program into manageable chunks.

- Functions may return a value, thus they have 'type', just as for simple variables. Unless specified otherwise, functions are assumed to return a value of type `int`.

- A value is returned from a function using the `return` statement.

- Values are passed into a function using parameters. Under ANSI C the type and name of these parameters are specified within the parentheses forming the function header. In classic C only the name of the parameters is contained within the parentheses; the type declarations are situated after the header line but before the body of the function.

- It is not a good idea to mix ANSI and classic C functions within the same program.

- The ANSI standard introduced function prototypes. These inform the compiler of the number and type of any parameters and also the function type. This enables the compiler to check that the function usage is consistent with the function definition.

Exercises

1   Enter the function given in Fig. 14.4 which raises an integer to a given power. Add a main program and function prototype and verify that it works correctly.

2   Modify your answer to the first exercise so that it raises a real number to an integer power. Do not forget to modify the function prototype.

3   Write a function which uses `printf` to print a given parameter value together with its square. The parameter should be of type `int`. Then modify the original squares program (see Fig. 11.1) to call this function.

15

An interlude

Aims and objectives

The objectives of this chapter are to:

- assimilate the knowledge so far;
- introduce the problem–solving technique used later in the projects.

Overview

This chapter is designed to take a break from learning grammar and provide a breath of fresh air by writing a simple program. The text follows the same form as for the projects covered later and should provide a useful end product.

The idea is to produce a visual display of how far a program has got with its processing. For example, suppose a program was designed to calculate the one thousandth prime number and used a technique which required calculating all those up to this value. The program would sit and think for some time and it would be difficult to know how far it had progressed. Many a time a working program has been terminated prematurely because it was thought to be taking too much time only to find out that it had almost finished, and was working correctly all along. One could write out all the intermediate numbers, or say which number was being calculated, but this project is to draw a simple graphic display.

Suggested output

The suggested output is as shown below:

```
>>--+----+----+----+----+----+
```

A single line is drawn consisting of − signs with a + sign every fifth character. A total of fifty characters are printed each of which represents two percent of the total job. A greater than sign indicates how far the program has progressed. This form of display will work only if there is no other screen output.

Note that the project is to handle the graphical display only, not to calculate prime numbers. An empty loop will be used to simulate the calculation.

Design

The graph is to be drawn using a single function that will need to be called several times. An initial call to this function will define what the limit of the graph will represent: for the prime number calculation this would be 1000. Subsequent calls to the function would say how far the calculation had progressed, for example, it had reached prime number 837. It is assumed that the calculation involved in the main program is far more significant than the overhead of calling the graph function. All values passed to the function should be of type int and therefore not exceed 32 767.

Starting off

Write a main program containing a for loop where the loop counter goes from 0 to 32 000 inclusive. Within this loop call the function pgraph specifying a single integer argument which is the loop index count. Write a dummy pgraph function that does nothing. Do not forget to prototype the function that will return nothing and take an int parameter. Run the program and debug it so far! It should take a second or two to run, depending on the hardware in use.

Initializing

Now the fun starts. The function has to be initialized by its first call, thus there has to be some variable, preferably local to the function, which indicates this is the first call. Here is a classic use for a static variable, as the value must be preserved across calls to the function. Define such a variable (called first) and preset its value to one (i.e. true). Add to the function an if test which checks first, and if its value is true (non-zero) writes out the parameter value and then sets first to zero. Modify the main program so that before the loop is executed, the function pgraph is called with the parameter value of 32 000. This will be the first call to the function and will define the number of times pgraph has to be called subsequently to produce a 100% graph.

Test this much and verify that the function output consists of only one value of 32 000, the initialization value.

Drawing the grid

Next modify `pgraph` so that on its first call the axis is drawn using a suitable `printf` statement. The axis lines should consist of fifty characters, `----+` repeated ten times. Use the special character `\r` at the start and end of the format string to ensure the axis starts at the left margin and to return the cursor to the start before processing begins.

Calculating the cursor position

After the first call to `pgraph`, each subsequent call will specify how much of the processing has been done. This will be a number in the range 0 to 32 000. This number has to be converted into a character position on the graph. For example, the value 16 000 should correspond to position 24 (i.e. 50% complete). To do this some form of scaling is required. The required position on the graph would be given by:

$$\text{position} = 50 \times \frac{\texttt{current parameter value}}{\texttt{maximum parameter value}}$$

The maximum parameter value is specified by the first call to `pgraph` while the current parameter value would change with each call. In order to calculate this expression the maximum value must be saved on the first call to `pgraph`. A few moments thought about the above formula will show that the intermediate result of fifty times the parameter could exceed the size of `int`, thus this expression must be evaluated using variables of type `long`. To avoid frequent casting it seems sensible to save the maximum parameter value in a variable of this type. Add a `static long` variable to `pgraph` which is set to the value of the parameter on the first call to the function. Do not forget to cast it correctly.

Next we need to add the processing required when it is not the first call to `pgraph`. This will require a calculation similar to the above to work out how far along the axis the pointer should be. This calculation will be something like:

```
col = (int) ((long) num * 50L / limit);
```

where:

```
col   = required column number for pointer
num   = parameter value (of type int)
limit = maximum parameter value (of type long)
```

The variable `col` now contains the required column number for the pointer.

Updating the screen

The problem now arises as to the current location of the pointer. If the pointer is already in the correct place there is nothing to write on the screen. Thus another static variable (pos) is required that contains the current position of the pointer on the axis. This should be initialized to zero, probably by the first call to pgraph. If the current pointer position (pos) is not equal to the required pointer position (col) then output the required number of greater than symbols and update pos accordingly. Or, to put it in a more C–like manner, while the current position (pos) is less than the required position (col) output a '>' and increment the current position.

To summarize: on the first call to pgraph clear the first call flag, clear the pointer column number, save the parameter value in a long variable and print the axis. On subsequent calls calculate the required column number for the pointer and update both the current position and the screen as required.

A possible solution is shown on the following page, but try to reach your own solution before studying it. It is easy to see how to solve a problem once one has seen the solution, it is far more useful to derive the solution independently.

Solution

```
/*   Simple progress graph routine   */

#include <stdio.h>

int main(void);
void pgraph(int);

main()
{
  int i, j;
  pgraph(32000);                      /*   Initialize   */

  for (i=0; i <= 32000; i++) {
    for (j=0; j < 10; j++)            /*   Extra delay   */
      ;
    pgraph(i);                        /*   Plot progress   */
  }
  return 0;
}

void pgraph(int num)                  /*   Update progress
{
  int i;
  static int first = 1;
  static int pos;
  static long limit;

/*   If first call initialize and print axis   */

  if (first) {
    first = 0;      /*   Clear first call flag   */
    pos = 0;
    limit = (long) num;
    printf("\r----+----+----+----+----+"
          "----+----+----+----+----+\r");
  }

/*   Otherwise update current position   */

  else {
    i = (int) ((long) num * 50L / limit);
    while (pos < i) {
      printf(">");
      pos++;
    }
  }
}
```

Comments on the solution given

The above solution is not ideal. A 'better' solution may be to have two separate functions, one to initialize the graph, and one to update it. A data structure would be used to pass the required values between the two functions. As suitable data structures have not been covered the other ways of keeping the values would be either in global variables or as function arguments.

Global variables should be kept to a minimum. They should be used only for a limited number of variables which are shared extensively between functions. Their meaning should be clear and routines that change them well documented. The above simple function does not—in the author's opinion—merit the use of a global variable.

Arguments could have been used. Questions of the inefficiency of passing many arguments can be ignored, updating the screen is very slow compared to passing a few arguments. This was not done as part of the idea of this project is to highlight how static variables can be used to retain values between calls to the same function. Also, if arguments are to be used, then these variables will need to be defined in the calling function. This requires more knowledge of the graph drawing routines.

The solution given using static variables requires the minimum amount of knowledge about how the function works. Only the function name and the permitted values for the argument need to be known for the function to be used.

It should be noted that the above solution requires no variable declarations outside of the function. It is totally self-contained. If a two–function approach were adopted then, as well as the two functions, one or more variables would need defining. This, the author feels, makes the incorporation into an existing program slightly harder, especially if there is a conflict of variable names.

A warning

When this solution was prepared the loop counter was set to 32 767 and not to 32 000 as above. The value of 32 767 was used as it is the largest signed number allowed in a 16 bit integer (the Microsoft C default). When the program was tried it appeared to work but after completing the bar graph the computer appeared to 'hang' and required rebooting to clear it. This is worth describing as it highlights what to watch out for when working at limiting values.

The error occurred with the `for` loop. The test for the loop being executed was that i `<= 32767`. This condition will always be satisfied because the value of i can never exceed 32 767 and so the `for` loop would never terminate. The moral of the story is to avoid the limiting values whenever possible, hence the use of 32 000 rather than 32 767 in the solution. This also emphasizes the need to test any program at the limiting values. Generally if these work, then the intermediate values will as well. Most compilers have options that enable checking for variables attempting to hold values outside their permitted range. Use of such options is recommended.

Extensions

1   Modify your program so that equals signs are used in the positions following the cursor. It looks nicer and a bit more professional. Use the special character \b to delete the pointer, then output => to update the pointer position. The output should be something like this:

```
=========>+----+----+----+----+----+----+----+----+
```

2   Normally only positive parameter values would be sensible. Modify the function pgraph so that if a negative value is specified the 'first call' variable is set back to a true value. Also output a newline. Thus if more than one progress graph is required in a program this can be accomplished by calling pgraph with a negative value once the first graph is complete and before the second one is required.

16

Pointers

Aims and objectives

The aims of this chapter are to:
- introduce pointer variables;
- give examples of their usage;
- define permitted pointer arithmetic;
- describe the void pointer type;
- describe the null pointer.

Overview

The display of an IBM/PC is memory mapped. In other words, there is a section of memory which is assigned to the display that the hardware continuously uses to update the display. This memory can also be updated by the processor chip. Thus to display something on the screen all one has to do is to place the required data in the correct bit of this display memory. This raises three questions: what format has the data to be in, at what address, and how? The last part is the subject of this section, the other two parts will be covered in the examples following.

Pointers

In order to save data at a given address there has to be a way of specifying 'an address'. To make it more general, a variable type which contains an address, rather than data values, would be preferred. In C terminology a variable which contains an address is known as a pointer. To complicate matters slightly the compiler needs to know the type of the data that the pointer will address.

In order to define that a variable is a pointer (i.e. that it contains an address rather than a data value), the variable name is preceded by an asterisk (*) in the variable type definition. Thus:

```
int  i;      i is an integer
int  *p;     p is a pointer to an integer
```

The idea of pointers is so much a part of C that it is worth restating. The unary operator '*' (which should not be confused with the binary multiply operator) means 'the contents of'. Thus int *p can be read as 'the contents of the address stored in variable p, are integer'.

There is a corresponding unary operator & which means 'the address of'. The expression &i gives the address in memory of where the variable i is stored. Do not confuse the unary & operator (meaning address of) with the binary & operator (meaning bitwise and).

As an example consider the diagram shown in Fig. 16.1. Each box represents a variable stored in memory. The value of the variable is contained within the box while the name is written to the right, all variables being of type int, except for p which is a pointer to an int. The memory address is written to the left of each box.

Fig. 16.1 Example memory diagram

100	3	i
102	5	j
104	−1	k
106	102	p

The value of variable i is 3 while its address is 100. In other words:

```
i is 3
&i is 100
```

Also, as p contains the value of &j then:

```
p is 102
*p is 5
```

The pointer operators can be used in expressions and have a higher precedence than most arithmetic, bitwise, or relational operators, but see Appendix B for details. The following is perfectly valid:

```
k = *p / 3 + 1;
```

The & operator will be used further in the second chapter on functions but work through the example shown in Fig. 16.2 and determine the final values of both a and b.

Fig. 16.2 Expressions using pointers

```
float  a=1.2F,  b=4.2F;
float  *p;
p  =  &a;
b  =  *p;
```

If you did not get both a and b equal to 1.2F then take a second look. The first assignment sets p to the address of the variable a. The second assignment sets b to the value of the variable whose address is stored in p (i.e. the value of a).

To summarize the two unary pointer operators are shown in Table 16.1.

Table 16.1 Pointer operators

Operator	Meaning
*	contents of
&	address of

Pointer arithmetic

Apart from using the value addressed by a pointer it is also possible to do limited arithmetic with pointers. Pointers can be incremented and decremented by any integer value. A pointer can be subtracted from another pointer providing both addresses are in the same memory block or array (more on this soon). Multiplication and division of pointers is definitely out: what would it mean anyway?

Consider incrementing a pointer by one. One what? is the next question. If the pointer were the address of a two–byte integer (for example the variable i in Fig 16.1), would the incremented pointer address the second half of the integer, or move on by two bytes and address what follows the integer? The second synopsis is logically correct and this is how C is defined.

A pointer is incremented or decremented in units equal to the size of the object to which it is pointing. This is another reason why the C compiler needs to know what type of variable a pointer is addressing.

Casting pointers

Where an expression includes pointers then, just as with simple variables, all the pointer types must be the same. Where pointers are different, casting will be required. Casting a pointer rarely changes the value of the pointer, it is after all just an address in memory, but it does enable the compiler to check for consistency of use. As an example:

```
int  i,  *p
char  *c;
p  =  &i;
c  =  (char  *)  p;
```

Void pointers

The pointer type void * is a special type of pointer introduced by the ANSI standard. A pointer of this type can be assigned to any other pointer type without casting. The most common use of void * is where a function returns a pointer and how this pointer is to be used is not known to the function. The memory allocation routine malloc (described further in Chapter 25) is a classic example of this. The function malloc returns the address of a block of memory that can subsequently be used to hold variables. As the type of these variables is not known to the function, malloc returns a void * pointer. Figure 16.3 shows its correct usage.

Fig. 16.3   Using void pointers

```
float  *p;
p  =  malloc(100  *  sizeof(float));
```

A pointer myth

There are two reasons why the code shown in Fig. 16.3 could result in the compiler issuing a warning message about incompatible types. Firstly the function prototype for malloc may have been omitted. In this case the return value would be assumed to be of type int and assigning this to a pointer would cause a compiler diagnostic message. Inserting a cast, as shown in Fig. 16.4, removes the error message, and will often work, but is not recommended. Often memory addresses can be stored in variables of type int but this is not guaranteed by the standard. Thus the assumed int return value from the function may have lost information that would not be recovered by using a cast. The correct remedy for the problem is to ensure the function is correctly prototyped.

Fig. 16.4   Superfluous casting under ANSI C

```
int  *p;
p  =  (int  *)  malloc(42  *  sizeof(int));
```

The second reason that the code shown in Fig. 16.3 could produce an error is if the compiler was not ANSI conformant. In pre-ANSI C `void *` did not exist. Functions returning generic pointers (e.g. `malloc`) used the type `char *`, thus an explicit cast, like that shown in Fig. 16.4, is required. The casting shown in Fig. 16.4 is unnecessary under ANSI C, but is required under the original C definition.

Null pointers

But what if an error occurs within the function `malloc`, or there is no memory available to allocate? In these cases the value NULL (defined in `stdio.h`) is returned. This value is distinguishable from an undefined value, a null value is defined and is known to point nowhere.

Within a program a null pointer can be specified either by using the constant zero or by using the symbol NULL. Just as with floating–point numbers, the internal representation of a null pointer need not be of concern. With either a zero constant or the NULL symbol the compiler still needs to be able to identify that the value represents a pointer. The following case shows where this is possible:

```
float *ptr;
ptr = 0;
```

However, if a pointer is used as an argument to a function which takes a variable number of parameters, then the compiler cannot detect that a zero value should be mapped into a null pointer. Defining such functions is beyond the scope of this text but such functions are available in the standard library.

As an example, let `fred` be a function taking a variable number of arguments, each of which is a pointer to a `char`. The end of the argument is indicated by a null pointer. A suitable call would be:

```
char i, j;
fred(&i, &j, (char *)0);
```

Finally do not confuse NULL with NUL. The latter is the name of an ASCII character with a zero value. An ASCII character NUL would normally be written as `'\0'`. Astute readers will have noticed that this is the character used to terminate character strings.

`near` and `far` pointers

Most compilers for the IBM/PC have an extension to the pointer definition. This is because the addressing system used by the processor has the concept of segments. An address can either be relative to the current segment (16 bits) or absolute (20 bits). To allow for this the keywords _near and _far are used.

A `_near` pointer contains a 16–bit address while a `_far` pointer contains a 32–bit address (a 16–bit segment number plus a 16–bit offset) that maps to the required 20–bit address used by the processor. Note that this extension is not part of the ANSI C definition although it is quite common on PCs. The keywords used are either `near` and `far` or `_near` and `_far`. Originally the names did not start with the underscore but this is now often added to conform with the ANSI C format for extensions. Keywords, macros, and functions supplied by the compiler vendor in addition to those required by the standard should start with an underscore character. Check your compiler manuals for the appropriate names. These keywords are used as follows:

```
int _near *a;
char _far *c;
```

One word of caution here. The `_near` and `_far` keywords bind to the variable name, not the type definition. Therefore in the following c is a far pointer but d is not.

```
char _far *c, *d;
```

With IBM/PC compilers there is usually a choice of memory models available. These define the type of addressing used within the program. For example the 'small' memory model allows the program to occupy one segment (64K bytes) and the data also up to one (different) segment. Thus all addressing can be handled using near pointers.

IBM/PC display memory

The `_near` and `_far` keywords have been introduced here because they are needed. Earlier it was stated that the display had its own area of memory. For text displays this starts at either address 0xB0000 or 0xB8000, depending on which display mode is in effect. The display mode determines the number of lines and columns on the display and, if appropriate, the number of colours available. Appendix C lists the available modes.

All text modes, except mode 7, use 0xB8000 so for this example assume this value and change it if necessary. The display memory contains two bytes for each character position on the screen. One byte contains the character while the other contains the character attribute. This attribute specifies foreground and background colours and whether the character is blinking or not. The two bytes are stored as character/attribute pairs for each display position starting from the top left corner sequentially along each line, starting a new line as required. The last position is the lower right corner.

Screen clear function

Ignoring the attribute byte, a simple function to clear the screen would be something like that shown in Fig. 16.5.

Fig. 16.5 Memory mapped screen clear function

```
void cls(void)
{
   int i;
   unsigned char _far *sptr;
   sptr = (unsigned char _far *) 0xB8000000;
   for (i=0; i < 80*25; i++) {
      *sptr = ' ';
      sptr += 2;
   }
}
```

There are two points to notice. First, a far pointer is used to specify the screen address. This allows the function to work irrespective of which memory model is being used by the compiler. This far pointer is preset using a 32–bit value, the top 16 bits being the segment number, the lower 16 bits being the offset in the segment. Using hexadecimal notation makes the separation between segment and offset more obvious. There are four bits per hex digit, thus the first four digits are the segment and the last four the offset. The 'real' address is the segment number multiplied by sixteen plus the offset. Secondly the unsigned keyword is used for the pointer since, although the character will usually be positive, the same need not be true of the attribute byte.

Enter the example program and verify that it works, modify the address to 0xB0000000 if it does not and try again. Also change the for loop terminating condition if you are using a display with other than 25 lines each of 80 characters.

The above program may fail because of an error like 'pointer range error', or words to that effect. This would be because the compiler is checking the value of every pointer before using it, and has detected that the pointer to the screen memory is outside the program space. This checking is not part of the standard but is frequently implemented as a debugging aid. With the option selected, a pointer is declared invalid if, when it is used, the value is outside the program's data area. Updating the screen memory would cause just such a situation. To prevent pointer checking there is usually an option or a menu item to turn it off.

Once working, modify the program to set the attribute byte to 0x07. Use an auto–increment of the form `*(sptr++)` to increment the pointer. Note that the brackets are required because without them `(*sptr)++` would be assumed. Refer to the operator hierarchy (Appendix B) and confirm this. `*(sptr++)` increments the pointer, `(*sptr)++` increments the value pointed to. Check that you follow this distinction; it is very important.

Summary

- Pointers are variables which hold memory addresses.

- There are two pointer operators: '`*`' meaning the contents of, and '`&`' meaning the address of.

- Pointers are incremented and decremented in units equal to the size of the objects to which they point. For example, on a system with four byte `floats` incrementing a pointer to a `float` will add four to the address.

- Pointers can be cast in the same manner as simple variables.

- There is a pointer type known as '`void *`' which is used as a generic pointer. A pointer of this type can be assigned to any other pointer type without an explicit cast.

- As with simple variables, the value of a pointer must be defined before it is used. Failure to do so may result in some arbitrary area of memory being used and/or changed.

- A pointer can contain an address that is known not to point anywhere. This value is known as NULL. A zero value, or the symbol NULL, can be assigned to a pointer to create this NULL pointer.

- C compilers for the IBM/PC allow the keywords `_near` and `_far` (`near` and `far` on earlier compilers) which allow the size of the pointer to be defined. `_near` corresponds to a two–byte address, `_far` to a four–byte address.

Exercises

1 Write a short program to output the current screen mode. This mode specifies whether text or graphics is being displayed, mono or colour, and the resolution. The various options are shown in Appendix C. The screen mode is saved as a single byte in segment 0x0040 at offset 0x0049. Your program should read this byte and then print out its value. Check the value is reasonable by comparison with Appendix C. Note that you should never attempt to change the screen mode by setting the byte at segment 0x0040, offset 0x0049. This byte should be treated as 'read only'.

2 Write a function to output a character and attribute pair at a specified point on the screen. This should be done using a function call of the form:

```
pch(sptr, col, row, c, a);
```

where:

```
sptr   = screen pointer (unsigned char _far *)
col    = display column, 0 to 79 (int)
row    = display row, 0 to 24 (int)
c      = character to write (unsigned char)
a      = attribute required (int)
```

The function should first check the values for the row and column are valid. Exit the function without further processing if either value is out of range. Then calculate the position in the display memory of the required character attribute pair. This is given by:

```
sptr + 2 * (row * 80 + col);
```

The factor of two allows for each display position having both character and attribute. The required values can then be saved as before, but do not forget to cast the integer attribute parameter to be an `unsigned char`. Write a main function that calls this function, initially to write just one or two characters. When happy all is working, modify the main function to fill the entire screen with letters using all the 256 possible attribute values. A `for` loop will be of use here. Persevere with this one as a modification to this function will be used extensively later. All the above figures assume an 80 character by 25 line display. Modify these limits if otherwise.

3 Having completed the previous exercise correctly try to display a table similar to Appendix A. This gives the characters displayed for each of the possible 256 character codes (0 to 255). Spend some time producing a neat table layout.

17

Arrays

Aims and objectives

The aims of this chapter are to:

- describe how to declare and use arrays;
- describe how arrays may be preset to given values;
- show their strong connection with pointers;
- define how to use multi-dimensional arrays.

Arrays

The description of arrays has deliberately been left so late in the text because the concept of pointers is so closely linked with them. Often, as will be seen shortly, the two are indistinguishable.

An array consists of a number of elements that can be referred to by using a single name and an offset. Each element of an array must be of the same type. In common with all variables, arrays must be declared before use and this is done as follows:

```
int matrix[10];
char string[30];
float data[100][10];
```

The number of elements in an array is specified within square brackets. Unlike many other languages the first element in an array is element zero (not element one) thus `matrix[0]` is the first and `matrix[9]` is the last integer value from the array `matrix`. An element of an array can be used anywhere a simple variable is valid.

The following program segment defines an array and sets the values of each element to the element's index (i.e. its offset from the start).

```
int i, num[6];
for (i=0; i<6; i++)
  num[i] = i;
```

Multi-dimensional arrays

Note the third example shown above which demonstrates how a two–dimensional array is declared. Both dimensions are enclosed in their own square brackets. Multi-dimensional arrays should be thought of as arrays of arrays, rather than as a single entity.

Should you ever wish to know, the last subscript varies fastest. In other words, the first three elements of the array data, defined as above, would be:

```
data[0][0]
data[0][1]
data[0][2]
```

The number of dimensions supported varies from compiler to compiler but need not cause concern because this is rarely a limitation and also because arrays of arrays (i.e. multi-dimensional arrays) are less frequently used than arrays of pointers.

Array initialization

An array can be defined and the values initialized in one statement as follows:

```
int num[6] = {0, 1, 2, 3, 4, 5};
```

Note that each entry in the initialization list must be a simple expression, usually a constant. This means that the values required can be determined during compilation. As with all initialization, values preset within a function that are not declared as static are initialized each time the function is called. Those defined outside of any function, or defined with the keyword static, are defined only once, when the program is loaded.

Just as with simple variables, arrays that are not preset are initialized to zero only if they are static, otherwise they contain undefined values.

The ANSI standard permits arrays to be preset as above whether the array is static or not. The original definition of C permitted arrays to be preset only if they were static. This meant that either the array had to be declared outside of any function, or the keyword static used. In either case the array was only initialized once, when the program was loaded.

If an array is preset then its dimension can be omitted. The compiler will allocate the required number of elements. This is often used when defining character arrays.

```
int num[] = {1, 3, 5, 7, 9};
char msg[] = "Illegal input";
```

String variables

The second example defines a character array, that is, each element can contain one character. The array is preset using a string constant which is a sequence of characters terminated by a zero byte. Using character arrays to hold strings is a common occurrence and these are often referred to as string variables.

Note how the braces may be omitted when presetting a character array. The array `msg` in the above example would be allocated at least 14 elements, 13 characters plus the terminating zero character. Remember that the end of a character string is marked by a zero value.

Large arrays

A word of warning about defining large arrays. Just as with simple variables, arrays defined within functions (including the main function) are dynamic, unless defined using the static keyword. Dynamic variables are placed in the stack memory which, on many systems, has a limited area. Under DOS the stack area is typically 8000 bytes. A dynamic array of 5000 integers would not fit and would result in a 'stack overflow' error message when the program was run. Large arrays should either be defined as static or, preferably, allocated at run time using the memory allocation routines. These are discussed further in Chapter 25.

Pointer and array equivalence

With three notable exceptions, when an array name is used in an expression the array name decays into a pointer to the first element of the array. All this rather confusing statement means is that when an array name is used in an expression then the value actually used is the address of the first element of the array. In other words if `num` is an array, then:

   `num` is treated as `&num[0]`

also

   `num+2` is treated as `&num[2]`

The direct extension of this is that:

   `*(num+2)` will be treated as `num[2]`

This equivalence will be described further in Chapter 19, but first the three exceptions.

Exception one is where an array is used as the argument to the `sizeof` function. Here the expression value is the size of the array, not the size of a pointer to that array. This is clearly what is usually required.

The second exception is where an array name is preceded by an ampersand (`&`) to obtain the address of the array. In the original definition of C this was illegal; however, some compilers did allow the syntax and simply ignored the `&`. In ANSI C an `&` before an array name prevents the automatic decaying of an array name into a pointer so that `&num` returns the address of the first element. But there is a difference between the type of pointer produced. If an array name decays to a pointer then the resulting type is 'pointer to type T' where T is the type of each member of the array. However if the `&` symbol is used before the array name then the address is of type 'pointer to array of type T'. The two types are different. The general rule is to avoid using the `&` before array names—it may not have the expected meaning—and beware of any code which does.

The third, and final, exception to the automatic decaying rule is when initializing string constants as in:

```
char s[] = "Hello World";
```

The text `Hello World` is an array of characters and, in an expression, would normally decay to the address of the first element. However, when presetting an array clearly it is the character values which are to be saved, not a pointer to them.

The above exceptions are a bit heavy but have been included for completeness. Many texts state that pointers and arrays are equivalent. They are not. A pointer contains an address while an array consists of a number of elements.

Pointers to strings

If one refers back to our very first program (Fig. 1.1) the function `printf` requires the address of a character string as its first parameter. If one were to use a character array instead, the format would be:

```
char msg[] = "Hello World\n";
printf(msg);
```

The above example defines a character array, presets its contents, and passes the address to the `printf` function for processing. The same end result can be produced by using a character pointer as follows:

```
char *msg = "Hello World\n";
printf(msg);
```

But the two forms are different. The first defines an array of at least thirteen elements (do not forget the terminating zero value) and passes the address of the first character to the function `printf`. On the other hand the second form again allocates at least thirteen elements for the text but also allocates a variable of type `char *` which is set to the address of the saved text.

Thus in the first example `msg` is an array which can be modified. In the second form `msg` is a pointer variable which contains the address of a text string. The text string is a constant and cannot be modified, although the pointer can be changed to address a different string.

Summary

- Arrays are declared in the variable type declaration statements by putting the required number of elements, within square brackets, after the variable name. For example the following declares `data` to be an array of 42 `int`s:

  ```
  int data[42];
  ```

- Array subscripts start at zero. Thus in the above example the 42 elements would be labelled `data[0]` to `data[41]`.

- For multi-dimensional arrays, each dimension must be enclosed within separate square brackets as in:

  ```
  int data2[12][4];
  ```

- If an array is preset then the dimension may be omitted.

- ANSI C allows all arrays to be preset, but Classic C only permits static arrays to be preset.

- Within an expression the name of an array decays to (i.e. is treated as if it were) the address of the first element of the array. There are three exceptions to this rule:

  (a) when the name is preceded by an ampersand (`&`);

  (b) when the array is the subject of the `sizeof` operator;

  (c) when a string constant is presetting a character array.

Exercises

1 Write a short program that includes a ten–element integer array. Each element of the array should either be preset, or set explicitly using a `for` or `while` loop, to a value equal to its index. In other words the first element is set to zero, the next to one, up to the last element which is set to nine.

2 Expand the previous program to print out the value in each element of the array and also the sum of all elements.

3 Write a program that contains a character array which is preset to some arbitrary text. The program should then write out the array one character at a time. Remember that the array will end with a zero value. To write a single character a suitable `printf` statement would be:

```
print ("%c", array[i]);
```

4 Again, expand the previous example but this time write out a full stop between each letter. For example `Paul` would become `P.a.u.l` (note there is no leading or trailing full stop).

18

Pre-processor

Aims and objectives

The aims of this chapter are to:
- describe the function of the C pre-processor;
- introduce the `#define` and `#include` keywords;
- show how macros are defined.

Overview

Within any program there are often limiting values. These may be maximum line lengths, or maximum values for particular variables. If these limits are used in many different places, and there is a requirement to change them, this could prove a lengthy job. Similarly having to prototype every standard function would be a bit tedious to say the least. To ease this problem all C source files are processed by a pre-processor before being passed to the compiler. The concept of a separate pre-processor program may now be a bit dated, as often the compiler also performs this function. The important thing is that what is compiled has passed through a processor and this is what is now covered.

The pre-processor is much like a text editor. It makes simple text insertion, substitution, and deletion. Commands to the pre-processor start with a hash sign (#), and do not require a trailing semi-colon. Spaces may precede the hash, but this is not recommended as some early pre-processors did not allow them. Also some pre-processors permit one or more spaces between the hash and the command but again this is not universal so don't use it. There are many commands available but only two will be covered here.

`#include`

The first common pre-processor command is `#include`. This reads the named file and inserts it into the source program at the current position. As a common example:

```
#include <stdio.h>
```

This should be found among the first statements of any C program and causes the file `stdio.h` to be read in. This file contains much useful information and function prototypes for performing input and output (`stdio` stands for 'standard input/output'). Among other things the function prototype for `printf` is defined here along with things like the name EOF. This is the value returned by input functions when the end of data is reached (EOF stands for End Of File). `stdio.h` is a normal text file and is worth listing as an example of an include file. It is usually found in a directory called `include`; for Microsoft C users the full directory path will be `\msc\include` if the default installation has been followed.

The delimiters around the include file name are significant. If, as shown above, angle brackets are used then the compiler supplied file is used. If double quote marks are used then the current directory (and possible other user defined directories) are searched for the file, and if not found then the compiler supplied files are checked.

When manuals are supplied with the compiler there is usually a definition of every supplied function and how it is used. Within this description it should specify which `include` file is required. For example the mathematical functions require `math.h`, while character string manipulation functions require that `string.h` be included. Failure to include the required files often results in function references producing warning messages due to missing prototypes.

Included files can themselves `#include` other files but often there is a limit. Ten nested includes is probably a safe limit but this is implementation dependent. Try not to include a file in itself!

#define

The `#define` command defines a character or character string to a name, for example:

```
#define MAXLEN 100
```

Everywhere within the C source the word MAXLEN will be replaced by the string 100. Therefore the input:

```
if (i > MAXLEN)
   printf("The value of i is too large\n");
```

would be converted by the pre-processor to:

```
if (i > 100)
   printf("The value of i is too large\n");
```

By convention the name used is upper case to avoid any clash with function names (which are generally lower case) and to highlight the fact that the name is 'defined'. Beware of using C format comments on #define directives as these comments will be taken as part of the character string and could cause problems when substituted.

Programmers familiar with Pascal may have noticed that where C uses { and } Pascal uses BEGIN and END. The following C program may look nicer to a Pascal programmer.

```
#define BEGIN {
#define END }
main()
BEGIN
    printf("Is this Pascal?\n");
END
```

Macros

Macros look very like function calls but the difference is that the macro is expanded into inline code by the pre-processor. This technique is often used where a small function would be required but performance requirements dictate that the overheads of doing a function call are too great. As an example of a macro, consider the following:

```
#define sqr(x,y)  ((x) * (x)) + ((y) * (y))
```

This defines the macro sqr that requires two parameters, x and y, and calculates the sum of their squares. To use this macro the statement would be:

```
z = sqr(alpha,beta);
```

As you can see, this looks exactly the same as a function call. Indeed many of the standard 'functions' are in fact macros.

The excessive number of parentheses in the macro definition are required. Do not forget that the pre-processor is simply a text substitution facility and knows nothing of mathematical hierarchy. If, for example, the parameters to sqr were 5 and a+3 consider what would be substituted with and without the parentheses. Without the parentheses the substitution would leave:

```
5 * 5 + a + 3 * a + 3
```

that is, $4a+28$, not the expected $a^2+6a+34$.

One further thing to be wary of is having an auto–increment (or decrement) on a macro parameter. Consider the result if the parameters to sqr had been 5 and a++, the auto–increment would be done twice, not once as one might have thought. This is seldom a problem but if anything like this is suspected then often the C compiler has an option that allows only the pre-processor to be run. The output from this is then available for examination.

Finally, do not leave a space between the macro name and the opening parenthesis. This could cause the pre-processor to miss the macro. An example of using a macro will be given in the project on finding a route through a maze.

Summary

- The pre-processor does elementary text processing on the C source before it is passed to the compiler.

- The `#include` directive inserts the contents of the specified file at the current position.

- The `#define` command defines a character string to a name. The name is usually in upper case.

- Macros can be defined for the pre-processor. These are defined by specifying the name and any parameters with parentheses. The pre-processor will recognise this syntax and substitute parameters.

- Beware of macro side effects such as auto–incrementing a parameter twice or getting the wrong evaluation order by not using sufficient parentheses.

Exercises

1   What is the difference between the following?

```
#include <stdio.h>
#include "stdio.h"
```

2   What does an include file contain?

3   Does a `#define` command require a semicolon as a termination character?

4   Write a macro definition to calculate the product of three numbers.

19

Functions (part two)

Aims and objectives

The aims of this chapter are to:
- introduce passing parameters by address;
- show how to return more than one value from a function;
- describe passing arrays to functions;
- describe passing multi-dimensional arrays to functions.

Introduction

The first chapter on functions described how values can be passed into a function using parameters and how functions can return a single value. The parameter passing method used is known as 'pass by value', as a copy of the original value is made available to the function. This chapter covers the alternative method of 'pass by address', which is also available with C.

Passing parameters by address

If a parameter is defined as being a pointer then this is known as pass by address. In effect a local variable is created that contains the address of the argument. This local variable is therefore a pointer and can be used to access and change the value in the calling function.

Consider the program segment shown in Fig. 19.1. In the main function the variable k is passed to the function sum as a pointer. Thus in the function sum the variable i is a pointer which contains the address of the variable k in the main function. When *i is used within the function sum, then the variable referenced is the variable k in the main function, not a copy of it.

Generally passing by address should be avoided where possible because passing by value, the alternative, tends to isolate functions more, and avoids variables being modified accidentally. Pass by address is needed when a function is required that returns more than one value.

Fig. 19.1   Parameter passing by address

```
main()
{
   int  k;
   ...
   sum(&k,  2);
   ...
}
sum(int  *i,  int  j)
{
   *i  =  *i  +  j;
}
```

Passing arrays to functions

If one considers how arrays could be passed to functions, the pass by value method can be seen to have a drawback. If an array containing 1000 elements was passed to a function using pass by value, then a duplicate array would need to be created and every element copied across. If the function only used one value out of the 1000 then the duplication would be very wasteful.

For this reason arrays are passed by address. If the name of an array is specified as an argument of a function, then the function is passed the address of the first element of that array. The concept of an array name decaying to a pointer was introduced in Chapter 17.

Study the function defined in Fig. 19.2, which is designed to calculate the length of a character string. Remember a character string ends with a zero value.

Fig. 19.2   String length calculation

```
int  strlen(char  str[])
{
   int  i;
   for  (i=0;  str[i]  !=  '\0';  i++)
      ;
   return  i;
}
```

The function expects to be passed an array and then uses a `for` loop to locate the terminating zero. The index of this position gives the number of characters in the string. Note the use of a null loop as all the required calculation is done within the `for` statement. Also the dimension of the array has not been specified. It could have been, but this would not serve any useful purpose other than possibly setting a maximum value if array subscript values were to be checked. The space for the array must be defined elsewhere before it can be passed to this function.

Figure 19.3 shows a very similar function that again calculates the length of a character string but this time uses a pointer variable as a parameter. Study the code, particularly the pointer increment, and check it makes sense.

Fig. 19.3   Alternative string length calculation

```
int  strlen(char  *str)
{
   int  i = 0;
   while  (*(str++)  != '\0')
      i++;
   return  i;
}
```

Having defined these two functions for calculating the length of a character string there are two questions to answer. How is each function called, and which is the better of the two?

The method of calling the functions is identical. Either function can be called as follows:

```
int  i;
char  name[] = "Fred Bloggs";
i = strlen(name);
```

Remember, an array name decays to the address of the first element in the array.

Which is the better function?

The first string length function (Fig. 19.2) has its parameter defined as being an array. Therefore it expects the address of the array to be passed. Effectively a local variable is defined in the function that contains the address of the array. But this is exactly how the second function (Fig. 19.3) is defined. The parameter is specified as being a pointer, whether it points to a single value or an array of values is not specified, just that it is a pointer.

Which of the two solutions is the better? In terms of readability it is whichever you prefer. Generally those who initially learn to program with a language that does not support pointers prefer to think in terms of arrays and thus the first solution is the more obvious. This is not wrong, but neater, easier to read solutions, often use pointers and familiarity with them is worth the time spent.

In terms of efficiency each time round the loop in the first example requires a calculation to find the value in the current array cell (`str[i]`) and an increment (i++). To get the value of an array cell requires first determining its address. This is done by multiplying the index value by the size of each array element and adding the address of the first element. Thus the whole loop requires two additions and one multiplication. The second solution only involves two additions, one to update the pointer (str++) and one to update the count (i++). Which is the more efficient will depend not only on the compiler but also on the actual hardware on which the program is run.

Defining array size

Passing arrays by address does raise one interesting point. How does the function know the size of the array?

For character arrays containing strings there is no problem. The end of the data is marked with a zero byte. But for all other arrays the function cannot tell the array's size. Either a special value needs saving by the programmer to indicate the end of data or, and this is probably better, a second parameter is passed which indicates the number of elements to be processed. Note that this value need not be the same as the array size in the calling function. For example in Fig. 19.4 the array data is allocated 100 elements but only the first 10 are used. This fact is passed to the function that prints the values.

Fig. 19.4 Passing array size to functions

```
int  i,  data[100];
for  (i=0;  i<10;  i++)
   data[i]  =  i;
pval(data,  10);
   .
   .
   .
void  pval(int  *p,  int  num)
{
   int  i;
   for  (i=0;  i<num;  i++)
     printf("data[%d]  =  %d\n",  i,  p[i]);
}
```

Note how the variable p is passed into the function as a pointer but is then used as if it were an array. As explained earlier there is no inconsistency.

If p is an array then, in an expression:

p[i] is equivalent to *(p+i)

If p is a pointer then, in an expression:

*(p+i) is equivalent to p[i]

Passing multi-dimensional arrays to functions

Passing a one–dimensional array to a function involves simply passing the address of the first element. For multi-dimensioned arrays (or, more accurately, arrays of arrays) the function needs to know the size of each dimension, although the size of the last dimension can be omitted if required. The example shown in Fig. 19.5 passes a two–dimensional array to a function.

Fig. 19.5   Passing a two–dimensional array to a function

```
int matrix[4][5];
clear(matrix);
    ...
void clear(int m[4][5]);
{
   int i, j;
   for (i=0; i<4; i++)
      for (j=0; j<5; j++)
         m[i][j] = 0;
}
```

The prototype for the above example needs some thought. For a one–dimensional array the name of the array decays to a pointer to the first element. This address is of type 'pointer to type T' where T is the type of each member of the array. For multi-dimensional arrays the array name decays to a pointer of type 'pointer to array of type T'.

For multi-dimensional arrays where each dimension is a known constant then probably the best way of defining the prototype is:

```
void clear(int m[4][5]);
```

The parameter name in the function prototype is ignored.

Passing variable multi-dimensional arrays

In the case where a function requires a parameter which is a multi-dimensional array, and the size of each dimension is not known at compile time, then the function requires the size of each dimension passed as a parameter and must do all the indexing internally. In this case pointer syntax is the norm.

Another point to note is the type of the parameter passed to the function. If a pointer syntax is required then the function must be passed a 'pointer to T' where T is the type of each element of the matrix. It should not be passed a pointer of type 'pointer to array'. Note the very different syntax used to pass the array to the function, and also the different function prototype: these are shown in Fig. 19.6.

Fig. 19.6 Passing variably dimensioned arrays as parameters

```
void clear(int *, int, int);

int matrix[4][5];
clear(&matrix[0][0], 4, 5);
   .
   .
   .
void clear(int *m, int x, int y)
{
   int i, j;
   for (i=0; i<x; i++)
      for (j=0; j<y; j++)
         *(m + i * y + j) = 0;
}
```

Summary

- If a pointer to a variable is passed to a function—rather than the value of the variable—then this is known as pass by address.

- When pass by address is used the function uses the actual variable in the calling function, not a copy of it. Thus functions using this method can effectively return more than one value.

- Arrays are always passed by address.

- When passing an array to a function either specify the length of the array or have some marker within the array showing where the end is.

- When passing multi-dimensional arrays to functions either explicitly declare each dimension of the array or pass the dimensions as parameters and do your own indexing within the function. This latter technique is required if the dimensions are not known until the program is run.

Exercises

1 Figure 19.1 shows a function for adding two numbers together where the first parameter is a pointer and the value is returned in this parameter. Enter this function, together with a main function to call it, and verify that it works as expected.

2 Write another function that adds two numbers together but this time the function value is the calculated sum. Thus neither parameter nor any variable used should be a pointer. Write a main function to call it and

check it all works. Compare this solution with that of the first exercise. Which is the easier to use? Do not use pointers just for the sake of it.

3 Rewrite the code shown in Fig. 19.2 to use a `while` loop and also change the code given in Fig. 19.3 to use a `for` loop.

4 Write a function to calculate the dot product of 2 two–dimensional matrices. The matrices are both of the form:

```
double a[2][2];
```

The function should take three parameters each of the above type. The first parameter is the returned value of the product of the other two. For those who need a refresher, the product of two matrices is given by:

```
A   B          a   b              A × a + B × c    A × b + B × d
         ⊗              =
C   D          c   d              C × a + D × c    C × b + D × d
```

20

Standard input/output

Aims and objectives

This chapter introduces the following topics:

- single character input and output;
- input buffering;
- formatted input and output;
- formatted input and output from and to character strings.

Introduction

One of the strengths of the C language is that it contains no input or output statements. While this might be considered a serious oversight this is not so. All C compilers are supplied with a standard library of functions, many of which are used for input or output. Although in the original definition of C the standard library contents were a bit vague, the ANSI standard now defines the minimum contents of the library.

To use the input and output routines supplied in the standard library one must include the header file `stdio.h`. This file defines the function prototypes of all the functions and also defines some symbols and some macros. One symbol defined is EOF. This is the value returned by an input routine when there is nothing left to read.

Character input

To read one character from the keyboard there is the function `getchar`. This requires no arguments and returns an `int` value which is the code for the entered character (normally ASCII). For example, entering the character 'A' would return a decimal value of 65 if using ASCII. The `getchar` function will wait until a key is pressed before returning the equivalent code. To indicate there is no more input (i.e. the End of File condition) enter control Z (the Control key and the Z key together). This will cause the `getchar` function to return a value equal to EOF, as is defined in `stdio.h`.

The necessity to return a value which is not a valid character is the reason getchar returns an int value. If the return type had been char then there would be no spare value that could be returned to indicate 'no more data'. The program shown in Fig. 20.1 demonstrates the getchar function by reading characters and printing their decimal values. These should match those given in Appendix A, if the ASCII code is used.

Fig. 20.1   Program to print character values

```
#include <stdio.h>

int main(void);

int main(void)
{
   int i;
   while ((i = getchar()) != EOF)
     printf("%c\t%3d\n", i, i);
   return 0;
}
```

Enter the above program and decide what it should do. Then try it and see if your prediction was correct.

Buffering

You probably expected that as each key was pressed, the key would be echoed, and its decimal equivalent given. Indeed, on some systems this would have happened, but on the DOS system the characters are only processed once an entire line has been entered. This is a feature of the operating system, not the C language, and occurs because some systems forward each character to the program as it is typed. Others, like DOS, work on a line–by–line basis. This process is known as buffering and should always be remembered when designing an interactive dialogue within a program.

If unbuffered single character input is required then this is often included as an addition to the standard library. The standard does permit such extensions.

Under DOS there is usually a function getch that behaves exactly as getchar except without any buffering. To use this function, the include file conio.h should be used.

Another useful feature is the ability to detect whether a key has been pressed. The technique varies with operating system, but with DOS a function kbhit exists that requires no arguments and returns a true value if a key has been pressed and the value not yet read. Normally a successful call to kbhit is followed by a call to getch. The include file conio.h is required to use this function.

Character output

The inverse of getchar is putchar. This function requires a single int value and writes the equivalent character to the screen. The putchar function also returns an int value. If all went well then this will be the same as the argument given. If there is an error and the character cannot be written then the value EOF is returned. The program shown in Fig. 20.2 copies what is entered at the keyboard to the screen. Before running the program predict its behaviour and then verify it. Do not forget buffering.

Fig. 20.2   A simple copy program

```
#include <stdio.h>

int main(void);

int main(void)
{
   int i;
   while ((i=getchar()) != EOF) {
      if (putchar(i) != i)
         break;
   }
   return 0;
}
```

Formatted output

Chapter 7 covered the printf statement in detail. Although here would be the correct place to define it, printf was covered earlier by necessity. The definition will not be repeated so refer to Chapter 7 if required.

Formatted input

The inverse function to printf is scanf. This reads input from the keyboard and converts the data according to a given format and puts the results in the specified variables. The format description is the same as for printf but with a few additions. The valid options are shown in Table 20.1.

Table 20.1   Format key letters (input)

Letter	Meaning
%c	Single character
%d	Decimal number
%i	Octal, decimal, or hexadecimal
%e	float in exponential or standard form
%f	Same as %e
%g	Same as %e
%o	Octal number
%s	Character string
%u	Unsigned decimal number
%x	Hexadecimal number

The input descriptors %e, %f, and %g are all treated similarly. Input can be in either standard or exponential form and the destination is of type float. To input a variable of type double the descriptors must be preceded by the letter 'l' as in %le, %lf, and %lg. For long doubles the additional letter is 'L' giving the descriptors %Le, %Lf, and %Lg.

The letter 'l' may prefix any integer key letter—as in %ld—to indicate the destination is of type long. Similarly the letter 'h' is used to indicate a destination variable of type short.

It must always be remembered that because scanf returns new values into the variables specified, it must be the addresses of the variables which are passed to the function. In other words the arguments must be passed as pointers. This is the reason the discussion on input and output was left so late.

For example, to read an integer value followed by a real value from the keyboard and save the results in the variables num and result the following command could be used:

```
scanf("%d%f", &num, &result);
```

Note how the variables are passed by address. The format descriptor (%d%f) specifies an integer followed by a real number. Any white space character or characters can separate these two numbers. In other words, any of the following lines would be valid input:

```
2  3.3
23 15.67
22                      3.14
```

It would also be valid to enter the two numbers on separate lines as in:

```
2
3.3
```

The format descriptor can contain any character sequence. Where they have no other special significance then the input should match these characters. For example, if the previous format descriptor had been %d, %f then the two numbers entered should be separated by a comma. If the data entered were not separated by a comma then the input would fail.

Talking of failed input, how does scanf return an error indication? As with most functions included in the standard library scanf returns a value. In this case the value is of type int and specifies the number of values correctly read. Thus if reading two values, as in the above examples, the scanf function should return a value of two. A zero value indicates no data were correctly read, and a value of one indicates only the first variable was correctly entered. If scanf returns the value EOF then either the end of the data has been reached, or an error has occurred, before any characters could be read.

The program given in Fig. 20.3 reads two real numbers from the keyboard and writes their sum and product.

Fig. 20.3 Simple input/output example

```
#include <stdio.h>

int main(void);

int main(void)
{
    float a, b;
    printf("Enter two numbers: ");
    if (scanf("%f%f", &a, &b) != 2)
        printf("Error entering data\n");
    else
        printf("Sum = %f, Product = %f\n",
               a+b, a*b);
    return 0;
}
```

Warning

As mentioned earlier, the function scanf requires the address of where to store the values, that is, a pointer. Failing to pass by address is probably the most common error associated with this function.

Another common error is to pass the correct type of arguments to the function, but not define valid values for these arguments. Consider the argument types for scanf when reading a single integer. The format string would be char * while the destination would be int *. The function prototype could therefore be thought of as:

```
int scanf(char *, int *);
```

At first glance, therefore, the following text might appear valid:

```
int *i;
scanf("%d", i);
```

While the parameters are all of the correct type for scanf, the program will probably fail. This is because although the variable i is of the correct type, it does not contain a valid value. It contains an address but unless the address is defined it will point anywhere. Note that the compiler will not detect this error as the program is syntactically correct. Just remember that defining a pointer variable does not automatically create a variable to which it points.

sprintf **and** sscanf

Finally two routines that, although not directly involved with input and output, fit nicely at this point. The function sprintf behaves in exactly the same manner as printf, but instead of the output going to the screen the output is written into a character array. This array is specified as the first argument. For example:

```
char line[10];
int i = 7;
sprintf(line, "TMP%03d", i);
```

The result of the above code would be to set the variable line to the character string TMP007. Code of this form could be used to create file names or format descriptors for printf statements.

As another example, suppose some repetitive calculation was being done when a hardware error was detected. Suppose also that there were a function available (which there is) called perror, which printed a given message and then followed it with a summary of the error. The following would provide a comprehensive error message:

```
char line[32];
sprintf(line, "Error during iteration %d", n);
perror(line);
```

The sprintf function returns a value of type int, which is the number of numbers converted. Be warned, however, that before the ANSI standard there were some implementations of C in which sprintf returned a value of type char *, this being the address of the buffer into which the text was written. When writing programs to be truly portable watch out for this one.

The inverse of `sprintf` is `sscanf`. This takes the specified character string and extracts values according to a given format in the same manner as `scanf`. As an example:

```
char buf[] = "1.2 3";
int i;
float j;
sscanf(buf, "%f%d", &j, &i);
```

The following chapter will introduce further uses for this function.

Summary

- If any input or output functions are used from then the file `stdio.h` should be included.

- `getchar` and `putchar` provide single character input and output. Both functions use and return variables of type `int`, not `char`.

- `scanf` and `printf` provide formatted input and output.

- `sscanf` and `sprintf` provide formatted reading from and writing to character strings.

Exercises

1 Write out the function prototypes for all the functions covered in this chapter. Note that for `printf`, `scanf`, `sprintf`, and `sscanf` a variable number of arguments can be specified. For this exercise assume a single integer variable is being read or written. The idea of a variable number of arguments is beyond the scope of this text.

2 Write a program to print out a list of names where the name starts with a preset character string (e.g. TMP) and ends with a numeric value. Ten names should be printed, the last two digits being a number in the range one to ten. Use `sprintf` to generate the names.

3 Modify the above program so that the characters which form the basis of the names are read in using `scanf`. To read a character string will require `"%s"` as the format descriptor.

4   Write a program that reads characters one at a time and counts the number of times each letter is entered. Ignore any characters that are not lower case letters and use a twenty–six–element array to store the counts. Note that it is perfectly valid to use character constants in expressions. For the purposes of this exercise assume the ASCII character set so that any character codes below ' a ' or above ' z ' should be ignored. The program should continue reading characters until an EOF is detected. Under a DOS system this is entered using control Z (i.e. the Ctrl key and the Z together).

21

File input/output

Aims and objectives

This chapter introduces the following:

- opening files;
- reading and writing files;
- closing files;
- file positioning;
- checking for errors.

Is it necessary?

Having mastered the standard input and output functions available for the keyboard and screen it is time to progress to file input and output.

Before embarking on file input or output, consider if it is actually necessary. For example, if a program is required to read from one data file and write the results to an output file, then it may be simpler to write the program to read from standard input (the keyboard) and write to standard output (the screen). This will be easier as no file actions are required. Once the program is written and debugged it can then be run using the DOS redirection operators to read from and write to files.

For example, if a program is written as described above it can be made to read from file `data`, and write to file `results` using the following:

```
prog <data >results
```

Choice of level

Having decided that file input/output is required there are two levels of file routines to choose from. One is a lower level 'efficient' method, while the other is easier and quite good enough for most use.

In case you are worried, only the easier routines are covered in this text: the lower level functions are more the preserve of system writers, or the enthusiastic! All the easy file access routines begin with the letter `f`. If a function appears to do what is required but the function name does not start with an f then ignore it, as the two sets of routines cannot be used interchangeably.

When using files there are three distinct phases required. The file must be opened, used, and closed. Each of these phases will be covered separately, the use being covered last.

Opening files

Before a file can be used it must be opened. This identifies the name of the file and how it is to be used. The function to do this is fopen and has the prototype:

```
FILE *fopen(char *, char *);
```

The function requires two character strings as arguments, and returns a pointer to something of type FILE.

The first character string is the external name of the file to be used. It can be a string constant or a variable as required, for example "file.dat". The filename can include a directory path but be careful; if using the backslash character in a string constant remember to escape it, thus:

```
"c:\\work\\file.dat"
```

The second character string determines how the file is to be used. The letters permitted are r (reading only), w (writing—any existing contents are lost), and a (appending—write after existing contents). Clearly only one of these characters can be used at a time. The two characters b and + may also be used in conjunction with r, w, and a.

The plus sign is a modifier to the r, w, and a letters. The combined forms meaning: r+ update mode (read and write), w+ update mode (read and write but delete any existing file contents), and a+ update mode (reading and writing but only writing at end of any existing data). Example calls are:

```
fopen("file.dat", "r");
fopen(fname, "w+");
fopen("users.dat", "a+");
```

A word of warning about 'a+' mode. The standard states that this mode does not allow any existing contents of the file to be overwritten. While this is generally how the classic C compilers worked, there are some systems that do permit the file contents to be overwritten.

Binary files

The letter b is used to indicate that the file is binary and is only required where the operating system uses more than one character to signify the end of a text line. Unix uses a single character (0x0A) but DOS uses the character pair 0x0D 0x0A. When reading a normal file, the DOS end of line character pairs are automatically converted to the single digit '\n' expected by C. Conversely when writing, the single character is converted to the character pair. However, if the file being read was data, for example a satellite image, it is possible that the character pair 0x0D 0x0A could occur and should not be converted to the single character form. This is where the letter b would be used to suppress this conversion. Note also that when a file is opened using binary mode then—when using DOS—the 'end of file' byte (decimal 26) is ignored.

Return value

As mentioned earlier fopen returns a pointer to something of type FILE. As yet structures have not been defined, but this is an example of one. The pointer returned by fopen is the address of block of memory, allocated by the system, which keeps track of one file. It could include, for example, the current position in the file and the last byte read. If the fopen call fails then a NULL value is returned. Always check for this, as it is pointless trying to use a file that has not been opened!

The returned value is used in subsequent file access functions as an identifier for the file. It is often called the file handle.

Predefined file handles

When a program is run three files are automatically opened. These are standard input (the keyboard), standard output, and standard error (both to the screen). These files can be used using the file handles stdin, stdout, and stderr. Sometimes it is beneficial to use a file input or output function, specifying one of the predefined file handles, rather than use the non-file function. An example will be given shortly when discussing error messages.

Closing files

Having finished with a file it should be closed using the fclose function. This causes any data in the file buffer to be written to the file and releases the memory used by the associated FILE structure. The fclose function requires a single parameter which is the file handle, that is, the pointer to the required file's FILE structure. As an example:

```
fclose(fh);
```

When a program terminates, any files that are still open are automatically closed. It is generally considered good practice to close all files that have been explicitly opened.

Some operating systems specify a limit of the number of files that can be open at once. For DOS this is usually in the region of twenty files. It is rare to need more files than this open at one time. Even if more files are required by the program it is usually possible to close one or more unwanted files to enable the new ones to be opened.

Character input/output to a file

Chapter 20 introduced the functions `getchar` and `putchar` to input and output a character from standard input or to standard output. The equivalent routines for file access are `fgetc` and `fputc`. In both cases one additional argument is required to specify the file. For example:

```
c = fgetc(fh);
fputc(c, fh);
```

`fgetc` returns the next character from the file or the value EOF (defined in `stdio.h`) if either the end of file is reached or an error has occurred. `fputc` writes the given character to the file and returns the character written or the value EOF if an error has occurred. The program shown in Fig. 21.1 copies the file `in.dat` to the file `out.dat`.

This example uses a standard function that has yet to be described. The function `exit` terminates a program. It can be used in any function and requires an argument of type `int`. A zero value indicates the program executed correctly while any other value indicates some form of error.

Figure 21.1 demonstrates the standard file handling procedure. Initially use `fopen` to open the required files and be sure to check the return values. Do not continue if the files are not opened. Then read from and write to the files using input and output functions with `f` as the initial letter. Finally close the files using `fclose`. Should the program fail then, in this example, the files are not closed explicitly but will be closed by the operating system.

Formatted file input and output

The equivalent routines to `printf` and `scanf` described in Chapters 7 and 20 are `fprintf` and `fscanf`. Again, one additional argument is required, the file handle, otherwise the functions are the same. For example:

```
fprintf(fh, "The value of i is %d\n", i);
fscanf(fh, "%d", &j);
```

As with `printf`, `fprintf` returns the number of characters output, or a negative value if an error occurred. Similarly the return value of `fscanf` matches that of `scanf`, it being the number of fields correctly converted, or the value EOF if no data are available.

Fig. 21.1   Simple copy program

```
/*   Simple  copy  program   */
/*   P.  Jarvis   28/01/91      */

#include <stdio.h>

int main(void);
void exit(int);

int main(void)
{
    int c;
    FILE *fin, *fout;

    fin = fopen("in.dat", "r");
    if (fin == NULL) {
        printf("Unable to open file in.dat\n");
        exit(1);
    }

    fout = fopen("out.dat", "w");
    if (fout == NULL) {
        printf("Unable to open file out.dat\n");
        exit(2);
    }

    while ((c=fgetc(fin)) != EOF) {
        if (fputc(c, fout) != c) {
            printf("Error writing out.dat\n");
            exit(3);
        }
    }
    fclose(fin);
    fclose(fout);
    return 0;
}
```

Try rewriting the example shown in Fig. 21.1 to read and write integers using %d format. Ensure the input file is in a suitable form and see what happens. Note that the output file may differ from the input file in layout if, for example, the input file contains several numbers on the same line. Do not forget to check the return values of the functions.

Error messages

As a matter of good form error messages should normally be written to the standard error file. Until now the default standard output has been used, but this is not a good idea if, for example, the output of a program were redirected into a file rather than to the screen. This feature is available on most operating systems, DOS included. If a program called `fred` writes ten numbers to the screen then it can be run using the following command:

```
fred >xx
```

As a result the ten numbers will be written to the file `xx`. Fine, but what if the program failed and tried to write an error message? If the message were written to standard output then it would be written to the file. Not much use there! By writing the error message to standard error instead, it would still appear on the screen rather than in the file. As an aside, it is possible on some systems to redirect standard error to a file, but that is going too far for now. Just note that the correct method of writing error messages is:

```
fprintf(stderr, "Error message\n");
```

String input and output to files

Although not mentioned in Chapter 20 there are two standard functions for reading and writing character strings from standard input and standard output. The omission was deliberate as the file routines should be used in preference. First a brief description of the non-file versions.

gets and puts

`gets` reads one line of input from the keyboard into a specified character array. For example:

```
char line[100];
gets(line);
```

If an error or end of file condition occurs then `gets` returns a NULL value. Otherwise it returns a pointer to the array passed as the argument. The drawback of this routine is that if a line is entered which is larger than the size of the array given, then the input will happily write over whatever follows the array.

The matching output function is `puts`. This writes the given character string to the screen and returns the value EOF if an error occurs. There are no nasty side effects with this function so it can be used freely.

fgets and fputs

The file equivalents to these routines are `fgets` and `fputs`. The `fgets` function takes three arguments. These are: the character array for input, the character array size, and the file handle. The array size must allow both for a new line character, and the terminating zero byte. For example:

```
char line[100];
fgets(line, 100, fh);
```

The specification of the input array size prevents the function writing past the end of the array but does introduce a slight catch. With `gets`, unless it failed, the input is always one line. There is therefore no requirement to save the end of line code in the buffer, so the character string ends with the '\0' character. With `fgets`, however, a long line could exceed the buffer length with the result that the first call would input the first part of the line and subsequent call(s) the remainder. Each time the function returned, the string read would be terminated by the '\0' character but on the last part of the line it would be terminated by both '\n' and '\0' (end of line and end of string delimiters). Like `gets`, `fgets` returns the address of the array or NULL if an error has occurred.

The `fgets` function can be used in place of the `gets` function using the predefined file handle `stdin`. This is always the preferred method though laziness often prevails.

```
fgets(line, 100, stdin);
```

`fputs` is the file equivalent to `puts`. It requires two arguments, a character string and a file handle. A return value of EOF indicates an error. For example:

```
fputs(line, fh);
```

The predefined file handle `stdout` can be used to make `fputs` act as `puts`.

Block file input and output

The functions `fread` and `fwrite` can be used to input and output a number of bytes to or from a file. No structure is implied in the data, it can be lines of text or part of an executable program. These functions are usually used to read and write binary data like satellite images and database files.

The `fread` function reads a block of data from a file. It requires four arguments as follows: a pointer to a buffer into which the data is to be put, the size of each item (in bytes), the number of items to read, and the file handle. Thus to read 2500 character variables would require:

```
unsigned char buf[2500];
fread(buf, 1, 2500, fh);
```

`fread` returns the number of objects read, which may be less than the number requested. The `feof` and `ferror` functions can be used to determine if the end of file has been reached or an error has occurred. These functions will be described later. The correct function prototype for `fread` is:

```
size_t fread(void *, size_t, size_t, FILE *);
```

The variable type `size_t` was mentioned briefly in Chapter 6. It is defined in `stdio.h`—usually as an `unsigned int`—and is the type returned by the `sizeof` operator. Generally either constants, or the result of the `sizeof` operator, are used as the arguments, thus casting is not required. For example to read 350 integers one would use:

```
int data[350];
if (fread(data, sizeof(int), 350, fh) != 350)
    fprintf(stderr, "Error reading data\n");
```

The inverse function is `fwrite`, which requires the same arguments and, believe it or not, in the same order. It returns the number of objects written which, if not the same as requested, indicates an error. The function prototype is the same as for `fread`.

`fseek` and `ftell`

These two functions are used to determine the current position within a file and to position the file at a given point.

`ftell` returns a `long int` value which is the current position in the given file. The value returned is the current offset from the start of the file. Usually this offset is measured in bytes but for true portability this should not be relied upon. The function prototype is:

```
long ftell(FILE *);
```

The function `fseek` is used to position the file pointer so that the next input or output function occurs at the specified position. The function requires the file handle, an offset value of type `long`, and a flag that indicates whether the value is an offset from the start of file, end of file, or relative to the current position. In order to rewind a file the following would be required:

```
fseek(fh, 0L, SEEK_SET);
```

The first argument is the file handle, then follows the numeric offset as a long integer, and finally the origin marker that can be one of three values:

`SEEK_SET`	Start of file
`SEEK_CUR`	Current position
`SEEK_END`	End of file

These functions will be demonstrated in the project on processing Tagged Image Format Files (see Chapter 35).

`feof` and `ferror`

These two functions are used when an input operation did not complete as expected. Both return either a true or false value. `feof` returns true if the end of data has been reached, `ferror` returns true if the last operation on the file failed for some reason. The normal usage would be:

```
if (ferror(fh)) {
   fprintf(stderr, "Error reading data file\n");
   exit(1);
}
```

`perror`

While talking about errors there is a useful function called `perror` that was briefly mentioned earlier. This function is used after an error has occurred to print out a message indicating the error. The function also requires a character string argument which is placed before the system error message.

Figure 21.2 shows a typical usage where the reason the `fopen` request failed is printed. Enter the program as shown and run it, ensuring that there is not a file called `nofile` in the current directory.

Fig. 21.2 Example use of `perror`

```
#include <stdio.h>

int main(void);
void exit(int);

main(void)
{
   FILE *fh;
   fh = fopen("nofile", "r");
   if (fh == NULL) {
     perror("error opening \"nofile\"");
     exit(1);
   }
   fclose(fh);
   return 0;
}
```

Summary

- There are two sets of file handling routines which cannot be used interchangeably. It is suggested one uses just those functions where the initial letter is 'f'.

- Files must be opened—using the fopen function—before they can be used.

- When they are finished with, it is good form to close the files, using the fclose function. Any open files will automatically be closed when the program terminates.

- Three file handles are automatically opened when a program is run. These are: stdin, stdout, and stderr.

- Error messages should be written to stderr, rather than the default stdout, so that the messages still appear on the screen even if the output has been redirected into a file.

- The input and output functions return a value if the functions fail. This value is usually EOF. This return value will occur when an attempt is made to read past the end of the data, and also if some error prevents the function completing. The functions feof and ferror enable the two events to be distinguished.

Exercises

1 Write a program to print the length of a file. The length can be determined by using the fseek function to go to the end of the file, and then using the ftell function to determine how many bytes the current position is from the start of the file. The file name can be given as a character string constant, as in Fig. 21.1 (but make sure the file exists). In fact Fig. 21.1 would be a good starting point for your solution. As an aside it is probably worth noting that the file offset does not have to be in bytes. It is on PCs and many other systems but is not defined as such in the standard.

2 Write a program that contains a string constant that would be a valid filename, but without a file extension. For example:

```
char name[] = "fname";
```

Then, using the sprintf function together with fopen, open two files one called fname.in and one called fname.out. Do not hard code the file names but build them from the base name using the sprintf function. When happy that this much works, extend the program so that any extension specified on the original file name is ignored. This sort of processing is often used, for example by compilers, where one source filename (e.g. `test.c`) implies some output file name (e.g. `test.exe`).

3 Write a program to count the number of lines in a text file. This can be done by reading the file one character at a time and counting the number of '\n' characters.

4 Modify the above solution (once working!) to print the length of each line. Do this by getting the current position in the file when a '\n' character is read. Use the `ftell` function to do this. The line length can be determined by subtracting the current value from the previous one (which should be initialized to zero). This is not the most efficient method but it demonstrates a point.

To get accurate offsets the file should be opened in binary mode so that the end of line code is counted as two bytes. In this exercise, however, the value required is the difference between two end of lines and hence the normal text mode is sufficient.

22

The main function

Aims and objectives

The intention of this chapter is to:

- describe how to return a value when the program terminates;
- define the meaning if this value;
- show the form of arguments passed to the main function.

Overview

As has already been stated, a function with the name main is required in all programs, and is where the program starts execution. The function main is the same as any other function except that it is called from, and returns to, the operating system, rather than another user–defined function. The parameters received by the main function, and the value returned from it, follow strict rules. The return value is the simpler so this is covered first.

exit

The function exit is used to return from any point in the program back to the operating system. Its use is not restricted to the main function. exit requires an integer argument, and its prototype—defined in stdlib.h—is as follows:

```
void exit(int);
```

The exit function terminates the program, closes any open files, and returns the given parameter value as the program completion code. How this is used is operating–system dependent, but a zero value is taken as meaning the program has run correctly. A non-zero value indicates the program failed in some way, possibly invalid input or an arithmetic error occurred. Whether this return value is subsequently checked is up to the user of the program.

return **within the main function**

In the main function—and only in this function—the return statement acts like a call to the exit function. Whether or not to use the exit function within main is purely a matter of style. The author prefers to use the exit function—with a non-zero argument—whenever the program fails, and use a return 0; to indicate successful completion.

If the main function completes without either the exit function being called, or a return statement being executed, then an arbitrary value is returned to the operating system. Readers using an environment version of C (e.g. Microsoft Quick C) will probably get a message when the program terminates similar to:

```
Program returned 42
```

All subsequent exercise solutions should include either call(s) to the function exit, or return statement(s), with the appropriate value(s).

The type of main

If the function main were terminated using a call to the function exit then logically the function would have type void. Therefore the program would end:

```
        .
        .
        exit(0);
}
```

and the corresponding prototype would be:

```
void main(.....);
```

However, the original definition of C defined main to return a value of type int (the default for all functions unless specified otherwise). The standard preserved this definition, maintaining the idea of a program's completion code being a 'return' value. Thus the correct prototype is:

```
int main(....);
```

If the exit function is used at the end of the main function then it should be followed by a return statement. The return statement will never be executed and is a waste of space, but without it the compiler would complain that the function did not return a value of the required type.

It is not unusual to see a void definition for the main function when the compiler allows it, but for strict conformance and hence better portability, the correct definition of type int should be used.

Parameters to main

The arguments supplied to the main function are supplied by the operating system. Thus, if they are to be used, the function must use the correct definition. When the program is run, any text in the command line that invoked the program is available through the parameters. For example, supposing the program was called copy and run using the command:

```
copy file1 file2
```

Then three values would be available to the program. Each of the three words in the command (i.e. copy, file1, and file2) would be available to the program as character strings.

Two arguments are supplied by the system. The first is of type int and defines the number of character strings available. The second is an array of character pointers, one for each argument supplied.

The term argument is used here to mean the value passed into the function. The word parameter is used to mean the variables defined within the function header that are set to these values. The story is somewhat complicated by some texts using the terms formal argument (for parameters), and actual argument (for arguments).

By convention the two parameters are called argc and argv, formal argument count and formal argument values respectively, but there is no requirement to stick to these names. Before looking at the function prototype consider the example program shown in Fig. 22.1.

Fig. 22.1 Parameters within the main function

```
/*   Program to print parameter values */
/*   P. Jarvis.                25/01/1991        */

#include <stdio.h>

int main(int argc, char *argv[])
{
   int i;
   for (i=0; i<argc; i++)
     printf("%s\n", argv[i]);
   return 0;
}
```

Figure 22.1 shows a program that prints every parameter supplied to the main function by the operating system. Type it in and verify that it works. Readers using an environment method of running their programs may need to delve into their reference manuals to discover how to supply arguments to the program when it is run. If the environment used does not offer this facility, then the program shown in Fig. 22.1 may need running outside the environment to demonstrate correct results.

The program contains a `for` statement that loops for each supplied value, the number of which is specified by `argc`. For each value the `printf` function is called to write it out as a character string (that is, the `%s` format).

The example program should have shown that the first parameter passed to the function `main` is the name of the program. This can be used to write error messages which include the program name or, less often, vary the program's behaviour depending on the name used to run it.

It should be noted that whatever is supplied as an argument, it is always passed to the main function as a character string. Thus if a numeric value is to be passed then the given character string has to be converted to a numeric value within the `main` function. There are standard functions (`atoi` and `atof`) to do this.

Prototype for `main`

Finally the function prototype for the main function. Whether or not this prototype is required is a matter of debate. One school of thought is that all functions should be prototyped, hence `main` should be as well. The other argument is that, as `main` is rarely called within a program, a prototype is superfluous. Whatever the pros and cons, the prototype is usually written as:

```
int main(int, char**);
```

The return value and first parameter look fine, but the second parameter type definition can look a bit confusing. It means that the second parameter is a pointer to a pointer to a `char`. This can be explained as follows.

The second parameter is an array, each element of which is a pointer to a character string (i.e. an array of characters). In an expression an array decays to a pointer to the first element thus `argv` is a pointer to an array of characters (the first character string). Thus `argv` is of type pointer to array of type `char`. But this is not the same as `char**` (a pointer to pointer to a char). In a function prototype and a function definition the types pointer to array and pointer to pointer are treated equivalently. If you are unsure on this don't lose any sleep over it as you can use the following if preferred:

```
int main(int argc, char *argv[]);
```

Any variable names supplied within a function prototype are ignored.

While confusion reigns, just a bit more for good measure. Consider the case where it is necessary to check the first letter of an argument passed into the main function. This can be done as follows:

```
if (argv[i][0] == 'a')
    statement;
```

As `argv[i]` is a pointer to the ith character string (i.e. a character array), then `argv[i][0]` is the first character in that string. This again demonstrates the strong links with pointers and arrays. An alternative form of the same command would be:

```
if (*argv[i] == 'a')
    statement;
```

Pre-ANSI syntax of `main`

Figure 22.2 shows the pre-ANSI equivalent of the program shown in Fig. 22.1. Note how only the parameter names are enclosed within the parentheses, the types of each parameter being defined between the header and the body of the function. Of course this is no different than any other pre-ANSI function definition but it is included here just to emphasize the common construct in older programs.

Fig. 22.2   Pre-ANSI form of the function `main`

```
/*   Program to print parameter values */
/*   Pre-ANSI version                   */
/*   P. Jarvis.            25/01/1991    */

#include <stdio.h>

int main(argc, argv)
int argc;
char *argv[];
{
   int i;
   for (i=0; i<argc; i++)
     printf("%s\n", argv[i]);
   return 0;
}
```

Summary

- The function `main` is of type `int`. The value returned is known as the program completion code and can be used by the operating system to check whether the program completed successfully.

- The program should be terminated either by a call to the function `exit`, or—within the function `main` only—a `return` statement. A zero return value indicates the program completed successfully.

- The main function can choose to ignore any parameters passed to it. In this case its function prototype would be:

```
int main(void);
```

- If any parameters are required then the operating system will supply two. The first being of type `int`, and the second an array of pointers to character strings. The prototype is thus:

```
int main(int, char**);
```

Exercises

1 Write a `main` function that requires a filename to be specified as an argument. Write a suitable diagnostic message if no filename, or more than one filename, is supplied. Do not forget to exit using a zero or non-zero value appropriate to the situation.

2 Extend the previous exercise so that the specified file is opened. Again, issue a diagnostic message if the file is not opened successfully. The file should be closed before the function exits.

3 Finally, extend the previous solution to use `fgetc` and `putchar` to read the file and write the contents to the screen. If `fputc` were to be used in place of `putchar` what would be the required file handle?

23

Structures

Aims and objectives

This chapter covers the following topics:

- structure definition;
- structure usage;
- structures as parameters within functions;
- presetting structures.

Structure definition

Structures—or records as they are known in other languages—are a method of grouping together several variables under a single name. For example, a date consists of a day, month, and year. The name date is used to mean all three elements. Such a structure can be defined as shown in Fig. 23.1.

Fig. 23.1 Example structure definition

```
struct date {
    int day;
    int month;
    int year;
} d1, d2;
```

This example declares the two variables d1 and d2 to be structures of type date. This structure is defined in the same statement as containing three integer values known as day, month, and year. The name, or tag, of a structure is optional, but if included then it is defined between the keyword struct and the opening brace. In the above example the tag is date. If a tag is defined then the definition of the structure and the declaration of the variables with that structure can be done using separate statements. This form is shown in Fig. 23.2.

Fig. 23.2 Separate structure and variable definition

```
struct date {
    int day;
    int month;
    int year;
};
struct date d1, d2;
```

The structure name must be unique and can therefore be used to relate the structure definition to the variable definition. This separate definition is usually preferred as it allows all structure definitions to be grouped at the start of a program, external to any function. Scope rules apply to structure definitions just as they do to variables.

To use the members of a structure one uses the '.' member operator:

```
d1.day    = day of week for date d1
d2.month  = month of year for date d2
```

Members of a structure can be used in exactly the same way as ordinary variables. Structure members can be used in expressions and assigned as normal. Note that the member operator has a very high precedence (see Appendix B) and thus parentheses are rarely required. The following are all valid uses of structure members:

```
d1.day    = d2.day * 3 - 1;
d1.day    = d2.day;
d1.year   = 1990;
```

Arithmetic using structures

Arithmetic operations cannot be performed between structures. Members of structures can be used, but not entire structures. In other words, it is not valid to enter d1 + d2. This is quite sensible really if one considers what it means to add two dates. How would the compiler know the rules for incrementing the month if the day value went past the end of a month? Indeed, what would define how long each month is?

Apart from using a member of a structure, the only other operation originally permitted on a structure was to take its address, as in:

```
&d1
```

Using a member of a structure and taking the address of a structure gave all the functionality required when the language was designed. It is probably a good idea to restrict one's own usage to these operations both for the sake of portability and, to a certain extent, efficiency.

The ANSI standard allows one structure to be assigned to another, providing both structures are of the same type. Thus if date d1 is defined then the assignment d2 = d1; is now permitted.

Passing structures as parameters

Another new feature introduced by ANSI C is that structures may be passed into a function by value. Thus each element of the structure is passed into the called function that clearly must be expecting the parameter to be a structure. The function shown in Fig. 23.3 would print the given date structure.

Fig. 23.3   Passing structures by value

```
void pdate(struct date d)
{
   printf("%02d/%02d/%d\n", d.day,
                    d.month, d.year);
}
```

With large structures passing every element by value would be slow so it is common to pass the address of the structure to the function. Indeed, this is all that was allowed in the original C definition. Thus if the address of the structure were to be passed, the above example would need changing to that shown in Fig. 23.4

Fig 23.4 Passing structures by address

```
void pdate(struct date *p)
{
   printf("%02d/%02d/%d\n", (*p).day,
                 (*p).month, (*p).year);
}
```

Note the syntax (*p).day, which is a bit messy to say the least. If p is a pointer to a structure then *p is the structure. To access a member of this structure requires the member operator but, as this has a higher precedence than the pointer operator, parentheses are required. Thus (*p).day is the member day in the structure pointed to by p. Not surprisingly there is a shorthand method of writing this, which is p->day, where the pointing symbol is made by combining – and >. The date printing function would therefore normally be written as shown in Fig. 23.5.

Fig. 23.5 The preferred syntax

```
void pdate(struct date *p)
{
    printf("%02d/%02d/%d\n", p->day,
                p->month, p->year);
}
```

Structures instead of multi-dimensioned arrays

As another example of structures consider storing a number of two–dimensional points. Each point would have an x and a y co-ordinate. Such data could be stored using a two–dimensional array where one dimension was used for the x co-ordinate the other for the y. Thus:

```
p[0][0]  =  x0
p[0][1]  =  y0
p[1][0]  =  x1
p[1][1]  =  y1
```

With structures the solution is more readable. Consider the following:

```
struct point {
    int x;
    int y;
};
```

If each element of array p was a structure of the above type, then the first co-ordinate in the array would be given by:

```
p[0].x  =  x0
p[0].y  =  y0
```

p[0] being the structure and p[0].x being the member x in that structure. The section of code shown in Fig. 23.6 reads from the keyboard a number of co-ordinate pairs and saves them in an array of structures. The advantage of the structure syntax is that there can be no confusion between which element holds the x co-ordinate. With a two–dimensional array there is the choice of which dimension and which value (0 or 1) represents each co-ordinate. Generally good use of structures improves a program's readability.

Fig. 23.6   Reading data into structures

```
struct point {
    int x, y;
};
struct point p[50];
char buf[100];
int i = 0;
while ((i < 50) && (gets(buf) != NULL)) {
    if (sscanf(buf, "%d%d", &p[i].x, &p[i].y) != 2)
        printf("Illegal input ignored\n");
    else
        i++;
}
```

Structures within structures

In order to define a rectangle parallel to an axis it is sufficient to define the location of just two, diagonally opposing corners. A structure to describe such a shape could be defined as follows:

```
struct rect {
    struct point top_left;
    struct point bottom_right;
};
```

The definition of one structure in terms of another is valid, providing all referenced structures have been previously defined and, obviously, a structure cannot form part of its own definition. The reason behind this is that the compiler needs to be able to determine the size of every element within a structure when that structure is defined.

In the above example, to refer to the x co-ordinate of the top left–hand corner will require two member operators, such as `fred.top_left.x`, where `fred` is a variable of type `struct rect`.

Pointers to structures within structures

It is possible, and very common, to define a pointer to a structure within the structure's definition. This is valid as, although the size of the structure is not known (as it is not yet defined), the size of a pointer to a structure is known. For example:

```
struct rect {
    struct point top_left;
    struct point bottom_right;
    struct rect *next;
};
```

ANSI differences

Just to recap, the ability to assign structures as in:

```
a = b;
```

where a and b are both structures of the same type, is new with ANSI C.

The passing of structures into a function by value is also new. Both new features can probably be ignored with no impact. Indeed passing structures by address is usually more efficient.

Presetting structures

Structures can be preset in a similar manner to arrays. Just as with arrays, in the original definition of C structures could only be preset if they were static. In ANSI C dynamic structures can be preset and are initialized each time the function in which they are defined is called. The following statements define and preset a variable using the earlier rectangle structure:

```
struct rect {
   struct point top_left;
   struct point bottom_right;
};
struct rect a = { {1, 2}, {3, 4} };
```

The inner braces are not required but help to emphasize that two internal structures are being preset.

Summary

- A structure enables a single name to encompass a number of elements. For example, a date structure could contain a day, month, and year.

- The member operator (.) is used to identify the required element within a structure. For example: d1.day.

- The operations permitted with a structure are:

  (a) a member of the structure can be used in an expression;

  (b) the address of the structure can be taken;

  (c) one structure can be assigned to another of the same type;

  (d) a structure can be passed by value to a function.

Exercises

1   Write a program which includes a date structure similar to that given in Fig. 23.2. Define one variable with this structure and preset each element to some suitable value. The program should then print out the date in the form dd/mm/yyyy (e.g. 15/10/1957).

2   Extend the above solution by adding a date input function. This should check that the values entered are legal (i.e. 30th February is definitely out), but ignore the effect of leap years for the moment.

3   Now add a date output function which should write a more readable form e.g. 14th March 1954. Do not forget to allow for the suffices st, nd, rd as in 1st, 2nd, 23rd etc.

4   A given year is a leap year if it divides exactly by four. However, it is not a leap year if it divides exactly by one hundred, but is if it divides exactly by four hundred. Write a function that, given a date structure, prints whether the year is a leap year or not. Expand this function to print the number of elapsed days to the start of the given year from the 1st January in the year zero. The formula for this would be:

```
y = year - 1;
days = y * 365 + y / 4 - y / 100 + y / 400;
```

In the above calculation make the variables y and days of type long as y * 365 could exceed the capacity of int. Note that any constants should therefore have the suffix 'L' to avoid mixing types (i.e. 365L rather than 365). Next add the number of elapsed days in the current year. This will require adding the number of days for each completed month (not forgetting to allow 29 days in February if the current year is a leap year) and the number of days in the current month. Finally, use the remainder operator (%) to determine the remainder having divided by seven. This will give a weekday number where zero is Sunday, one is Monday, and so on. Complete the function by making it print the day of the week, previous output can be removed, and verify using some known dates that the function works.

The formula used is based on the Gregorian calendar which came into use at different times. In the Western world September 1752 is the starting date so any earlier dates will not give real answers.

Also note that the number of days in year zero are used as part of the calculation. There never was a year zero (1 B.C. was followed by 1 A.D.) but assuming there was, does simplify the calculation.

24

Unions

Aims and objectives

The aims of this chapter are to:

- define the concept of a union;
- show how they can be used;
- demonstrate how to preset them.

Union definition

Unions are a rather strange idea, usually left as the preserve of systems programmers. However, they are covered in this text, both for completeness, and because they are used as part of the interface to the DOS operating system.

A union is defined in exactly the same manner as for a structure except that the keyword struct is replaced by union. Figure 24.1 shows a typical union definition.

Fig. 24.1 Union definition

```
union fred {
   int x;
   int y;
   float z;
};
union fred a, *b;
```

The usage of unions follows that of structures. The member operator . is used to refer to a specific member of a union and the -> operator can be used with pointers to unions. For example, given the above union definition, the following are valid members:

```
a.z
b->x
```

If unions are so similar to structures then what is the difference? The answer lies in how the various members are stored in memory. For a structure each member is distinct, the overall size of a structure being at least the sum of the sizes of the constituent members. On the other hand, all members of a union occupy the same memory locations, the overall size of a union being that of its largest member. For the example shown in Fig. 24.1 the union size is equal to four bytes (on a DOS system) as this is the size of `float`.

Thus in this example the variables `a.x` and `a.y` are the same. They occupy the same memory so that any change to one automatically is reflected in the other. If variable `a.z` is set then the values of `a.x` and `a.y` will be changed.

Using unions

The IBM/PC and clones use one of the 80x86 family of processor chips, the internal workings of which can remain a mystery. However, it is possible, from a C function, to ask the DOS operating system for some specific action; for example, to read data from a disk or to return the current screen resolution. The action required is specified by setting certain registers within the processor.

The names of the registers most often used are `ax`, `bx`, `cx`, and `dx`. Each can be thought of as a sixteen–bit unsigned integer variable within the processor chip. Each sixteen–bit register can also be referenced as two eight–bit values, a high and a low byte. Thus `ax` is composed of `ah` and `al` and similarly for `bx`, `cx`, and `dx`. If the value of `ah` is changed then the value of `ax` is also changed accordingly. Consider the definitions in Fig. 24.2.

Fig. 24.2 Example union

```
struct  WORDREGS  {
                    unsigned  int  ax;
                    unsigned  int  bx;
                    unsigned  int  cx;
                    unsigned  int  dx;
                  };
struct  BYTEREGS  {
                    unsigned  char  al,  ah;
                    unsigned  char  bl,  bh;
                    unsigned  char  cl,  ch;
                    unsigned  char  dl,  dh;
                  };
union  REGS  {
                struct  WORDREGS  x;
                struct  BYTEREGS  h;
              };
union  REGS  fred;
```

The union REGS contains two structures. The first contains four sixteen–bit variables, the second eight eight–bit variables. As it is a union the two structures overlap and thus `ah` and `al` together occupy the same two bytes as `ax`, and so on. The order of `ah` and `al` might appear strange but that is how the processor was designed.

How C programs can interface directly to DOS using this mechanism will be covered later, in Chapter 28.

Another use for unions is to store one of a number of different types of variable where the variable type is not known until the program is run. Consider the case where a value is to be entered and it can either be of type `int` or `float`. A union could be defined to hold this value but a flag would also be needed to indicate the valid type. For example, consider the following construct:

```
union value {
   int  i;
   float  f;
};

struct data {
   int  type;
   union  value  v;
};

struct  data  p1;
```

The structure `data` contains an integer type flag that indicates which of the union members is valid. For example, a zero value could indicate `int` while a value of one indicates `float`. Therefore if `p1.type` is zero then `p1.v.i` contains a valid integer, otherwise `p1.v.f` contains a valid floating–point value.

Presetting unions

The original definition of C did not permit unions to be preset. The ANSI standard now permits the first, and only the first, member of a union to be preset. This member can be a simple variable or a structure. For example:

```
union num {
   int  x;
   float  y;
};
union  num  var = 1;
```

Summary

- Unions follow the same rules as structures.

- The difference between a union and a structure is that in a union all the members occupy the same memory area, that is, overlap.

- Only the first member of a union may be preset.

Exercises

1   Write a program that includes a union definition as shown in Fig. 24.1. For variable `a` set member `z` and then print out members `x` and `y`. Then set `x` and print `y` and `z`. Does it make sense?

2   The `sizeof` operator can be used to determine the size of a union just as it can for a structure. Modify the solution to the previous exercise so that it prints the size of the union. Is this what you would expect?

3   Write a similar program but this time use the union definition as shown in Fig. 24.2. Set and print some of the values but note that to refer to members of the unions will require two member operators something like: `fred.x.ax` or `fred.h.al`.

25

Dynamic memory allocation

Aims and objectives

The aims of this chapter are to:

- describe the memory allocation facility available in the standard library;
- show how this facility is used.

Overview

Dynamic memory allocation is not a feature of the C language; rather, it is a facility offered by the standard library. Consider the problem of writing a program that requires a large array to store data where the size of this array is dependent on the amount of data to be processed. It would be possible to allocate a fixed size array when the program was compiled and let the program complain if too much data are entered. However, this system has its drawbacks. Firstly, if the program were ever moved to another machine the available space may be different requiring the array size to be modified. Secondly, on systems where jobs are charged real money the cost is often dependent on both processor time and on memory used, often the product of the two. Thus any waste of memory is a waste of money.

What would be nice is a mechanism for telling the program: 'an array of this size is required now, please may I have one?'. A request to release this memory, when finished with, would also be useful. These are exactly the facilities offered by the dynamic memory allocation routines.

`malloc`

There are several routines to request (or allocate) memory but for this text only the function `malloc` will be described. This function requests a block of memory and, if successful, returns the address of the allocated block. Figure 25.1 shows an example call.

Fig. 25.1   Example use of `malloc`

```
int *ptr;
int num = 250;
ptr = malloc(num * sizeof(int));
if (ptr == NULL) {
    fprintf(stderr, "Unable to allocate memory\n");
    exit(1);
}
```

The example demonstrates a number of points. Firstly `malloc` requires as an argument the number of bytes to be allocated. The `sizeof` operator is used to specify the number of bytes required for a particular variable type which then makes the program more portable. Thus for a DOS machine using two–byte integers the `malloc` function as shown in Fig. 25.1 would request 500 bytes.

Secondly `malloc` returns a pointer, this being the address of the first byte allocated. But `malloc` has no way of knowing how the memory is to be used. It could be for integers or doubles; it does not know. Therefore `malloc` returns a pointer of type `void *`, that is, here is an address but `malloc` has no idea how it is to be used. The pointer need not be cast because, as explained earlier, a `void *` pointer can be assigned to any other pointer type without casting.

The ANSI standard introduced the concept of `void *`. Prior to this `malloc` returned a pointer of type `char *`. This meant that often the pointer required casting (i.e. whenever it was not being used as a character pointer). Under the new definition, casting is never required.

Lastly, and probably most importantly, the return value from `malloc` is checked. As it is a pointer the returned value is checked for NULL. A NULL return value would indicate that the request for memory had been refused, probably because there was not sufficient memory left to honour the request. The function prototype for `malloc` is:

```
void *malloc(size_t);
```

Note that the number of bytes is specified using the system–dependent type `size_t`. For a DOS system, `size_t` is usually defined as an `unsigned int` thus the largest block of memory assignable in one go is 65 535 bytes. Other routines in the memory allocation family allow larger amounts but do remember this limitation.

There is an include file required for the memory routines and this is `stdlib.h`, but this is a change introduced by the ANSI standard. Prior to the standard, the include file required was `malloc.h`, and some implementations still use this. The only way to be sure is to check the manuals for the product being used.

Having allocated the memory it can be used in exactly the same way as any other pointer. For example, the code shown in Fig. 25.2 would initialize the entire array to zero.

Fig. 25.2 Presetting the allocated memory

```
int i;
for (i=0; i<num; i++)
    ptr[i] = 0;
```

free

Once allocated, an area of memory can be released back to the operating system using the function `free`. This requires the address of the block to be released. It is very important that this function is only used to release memory that has been requested using the allocation routines. Also it is invalid to try to release anything other than an entire block. Memory must be released in the same blocks as it was allocated, although the order of release is not important. The memory allocated using the code shown in Fig. 25.1 could be released using:

```
free(ptr);
```

Although it is good practice to return all allocated memory when it is finished with, the program will clear everything left when it terminates.

Summary

- Large arrays, and arrays whose size is not known at compile time, are best allocated at run time using the standard function `malloc`—or an associated function.

- The value returned by `malloc` is of type `void *` and is the address of the memory assigned.

- A NULL return value indicates that the memory was not allocated. Always check for this return condition.

- The allocated memory can be used in just the same manner as a variable or array declared at compile time.

- When the memory is no longer required it can be released using the `free` function.

Exercises

1   Write a program and incorporate the code shown in Fig. 25.1. Add a call to `free` to release the code and check it works. Try using the memory allocated by initializing all variables to their position in the array, that is, `ptr[0]=0`, `ptr[1]=1`, `ptr[2]=2`, . . ., and then print out their sum.

2   Modify the above program to keep allocating blocks, without releasing them, until `malloc` returns a NULL value. When this happens print out the total amount of memory allocated. The same pointer can be used for each call to `malloc` as the memory assigned will not be used.

26

The standard library

Aims and objectives

The aims of this chapter are to:
- summarize the major components of the standard library;
- describe how they are used.

Overview

Chapters 20 and 21 introduced the input and output available using the standard library while Chapter 25 introduced two of the memory allocation routines. The library contains many more useful functions, only some of which will be described here.

The documentation included with any C compiler should contain a library reference manual. This will describe all routines in detail and often contains a section describing the functions by topic. For example there are the maths routines, the string manipulation routines, and the character typing functions. Most implementations for IBM/PCs—or compatibles—include a graphics section, although this does not form part of the ANSI standard.

Maths routines

The Fortran programming language is often thought of as the language for processing mathematical functions. However, there is no reason why C should not be used for this. Indeed when the author needed to benchmark some new hardware a prime number calculation program written in C produced identical timings to the one written in Fortran.

There are numerous maths functions available, some of which are listed in Table 26.1, but note that these functions process and return values of type double, not float. This is a common source of errors initially but is quite sensible really. If one is processing mathematical variables it makes sense to use the maximum precision available, remembering that long double is a recent introduction and was not defined when the initial standard library was specified.

To use the maths functions from the standard library the file `math.h` should be included. This contains all the required function prototypes, together with some symbol and macro definitions.

Table 26.1 Some maths functions

Function	Usage
`sin(x)`	sine of x
`cos(x)`	cosine of x
`tan(x)`	tangent of x
`atan2(y,x)`	$\tan^{-1}(y/x)$ in range $-\pi$ to $+\pi$
`log(x)`	natural logarithm $(x > 0)$
`pow(x,y)`	raise x to the power y

As an example, the program shown in Fig. 26.1 produces a table of powers of two, from zero to ten. Note how the format string in the `printf` statement specifies a precision of zero, thus producing integer output.

Fig. 26.1 Program to produce a power of two table

```
/*   Program to print powers of two    */
/*   P. Jarvis.              28/02/1991 */

#include <stdio.h>
#include <math.h>
int main(void);

int main(void)
{
   double x;
   for (x=0.0; x<10.0; x++)
     printf("%3.0f - %4.0f\n", x, pow(2.0, x));
   return 0;
}
```

String functions

In C there are no string operators. In some languages strings can be concatenated, or joined, using the addition operator (+). This is not so in C, but the same—if not more—functionality is available using the standard library.

There are functions for joining, comparing, and searching for specific characters within strings, to name just a few. Again not all the available functions are listed, but Table 26.2 lists the more common. To use these functions the file `string.h` must be included. Warning: some early C implementations used the file `strings.h`, but this is not standard.

Table 26.2 Some string functions

Function	Usage
`strcpy(s1, s2)`	Copy string `s2` to string `s1`
`strcmp(s1, s2)`	Compare strings `s1` and `s2` (returns zero if strings are equal)
`strcat(s1, s2)`	Join string `s2` to end of `s1`
`strlen(s)`	Return length of string `s`
`strchr(s, c)`	Locate character `c` in string `s` (Returns character address or `null`)

Consider the program shown in Fig. 26.2, which reads lines from the keyboard and writes their length. The program terminates on either a blank line or an end of file character being entered. The use of the function `fgets` prevents the input buffer overflowing: lines longer than the buffer length will be counted as two separate lines. Note the `-1` in the string length calculation as the string will include the `\n` character.

Fig. 26.2 Program to print line lengths

```
/*   Program to print line lengths    */
/*   P. Jarvis.             28/02/1991 */

#include <stdio.h>
#include <string.h>
int main(void);


int main(void)
{
   int i;
   char line[100];
   while (fgets(line, 100, stdin) != NULL) {
      i = strlen(line) - 1;
      if (i <= 0)
         break;
      printf("Length = %d\n", i);
   }
   return 0;
}
```

Character class tests

Functions in this category determine the type of a given character. For example, is it numeric, or an upper case character? They should always be used in preference to a test for a numeric value as not all computers use the ASCII character set and programs should be portable whenever possible. It is also highly likely that the supplied routines are far more efficient than any 'home brew' equivalent.

To use these functions the file `ctype.h` must be included. Many of the supplied routines are macros, rather than functions, and are defined in this include file. Some common examples are listed in Table 26.3. Each of the functions requires and returns a value of type `int`.

Table 26.3 Some character class routines

Function	Usage
`isdigit(c)`	Decimal digit (0 – 9)
`isalpha(c)`	Upper or lower case character
`islower(c)`	Lower case character
`isupper(c)`	Upper case character
`ispunct(c)`	Punctuation character
`isspace(c)`	White space character
`toupper(c)`	Returns `c` as upper case
`tolower(c)`	Returns `c` as lower case

The section of program shown in Fig. 26.3 shows the `toupper` function being used to enable the case of an entered character to be ignored. It does not matter if the character is already upper case, or not even alphabetic, as the case conversion will only be done if the given character is lower case.

Fig. 26.3 Checking character input ignoring case

```
if (toupper(c) == 'Y') {
   ...
}
```

Warning: pre-ANSI the action of `tolower` was not defined if the argument was not an upper case character. Similarly `toupper` was only guaranteed for lower case arguments.

Conversion routines

There are also a number of useful routines for converting strings to numeric values. These require the include file `stdlib.h`. Some of the more common ones are listed in Table 26.4. Be warned: when a function uses a floating point variable it is usually of type `double`, not `float`.

Table 26.4 Conversion routines

Function	Usage
atof(s)	Convert string s to double
atoi(s)	Convert string s to int
atol(s)	Convert string s to long
abs(n)	Absolute value of int n
fabs(f)	Absolute value of double f
labs(l)	Absolute value of long double l

Note that, although the function `fabs` would logically fit with the other conversion routines, it is in fact part of the maths library and therefore requires the include file `math.h`.

A very common usage for these functions is extracting the numeric value of a character string passed into the program using parameters. Figure 26.4 shows such usage.

Fig. 26.4 Converting numeric parameters in the main function

```
#include <stdio.h>
#include <stdlib.h>
int main(int, char**);

int main(int argc, char *argv[])
{
   int n;
   if (argc != 2) {
     fprintf(stderr, "Requires one number\n");
     exit(1);
   }
   n = atoi(argv[1]);
   if (n == 0) {
     fprintf(stderr, "Illegal parameter\n");
    exit(2);
   }
   ...
}
```

Other functions

Not all the available functions have been listed. The projects covered later do not require any functions that have not been listed here, but one is free to use others if desired. As a general rule, check the standard library before writing a function if there is any possibility it might be a general requirement. The standard library routines should be optimized for the particular machine and should handle all the error conditions correctly. The library reference manual supplied with the compiler (or other reference style text) should be consulted for a detailed explanation of a function.

Summary

- The standard library is a good source of general purpose functions.

- Functions are grouped and there is usually an include file for each group.

- Standard library functions return values to indicate whether or not they completed correctly. These values should be checked.

Exercises

1   Modify the program shown in Fig. 26.2 to cope with lines longer than the length of the input buffer. This will require a check to see if each line read is complete. If it is, the last character in the line will be the newline character \n. This can be checked for once the string length has been determined. If the line is not complete then the length should be saved so that it can be added to the length of the following line. Do not forget to allow for one line being so long that it requires more than two calls to fgets to read it in completely.

2   Write a function that takes a character string as an argument. This string is assumed to be a question requiring a yes or no answer. The function should write the question and then wait for a reply. Decide whether the input has to be the entire word or just the first character, the latter being the easier. Read in the reply and return a true value if the reply is yes and a false value if no. If the reply is invalid write out a message to this effect and prompt for more input. If well written this function will be very useful.

27

Recursion

Aims and objectives

The aims of this chapter are to:

* describe the recursive programming technique;
* demonstrate how C can handle this construct.

Overview

Recursion is not part of the C language. Rather, it is a useful technique available to any stack–based language, of which C is one. Recursion involves a function calling itself. While this may seem strange, providing two simple rules are followed, there is no hidden magic.

The first rule for recursive algorithms is that the problem can be simplified in terms of itself. The factorial problem is the perennial example of this, so why be different? Factorial n (written n!) can be defined as:

```
n! = n × (n-1)!
```

The second rule is there has to be some condition that indicates the end of the required recursion. In the factorial example the terminating condition is that factorial zero is one, that is, $0! \equiv 1$.

The example shown in Fig. 27.1 shows a function to calculate a given factorial. The parameter passed is assumed to be positive.

Fig. 27.1 Factorial calculation

```
int fact(int n)        /*  Calculate factorial n  */
{
   if (n == 0)
      return 1;
   return n * fact(n-1);
}
```

Enter this function together with a main function to call it and verify that it works. Beware of using too large a value as the variables used are only of type `int`.

If unsure insert a `printf` to print out the value of the parameter passed, or alternatively enter the function shown in Fig. 27.2 which does much the same, but includes a few diagnostic messages.

Fig. 27.2   Diagnostic version of the factorial function

```
int fact(int n)         /*   Diagnostic factorial   */
{
   printf("Entering value - %d\n", n);
   if (n == 0)
      n = 1;
   else
      n = n * fact(n-1);
   printf("Leaving value - %d\n", n);
   return n;
}
```

Another mathematical calculation which lends itself to a recursive solution is that of calculating the greatest common denominator (gcd) of a pair of numbers. The gcd of two numbers is defined as the largest integer number that divides exactly into both numbers. The calculation involved is as follows. If both numbers are the same then either value is the gcd, otherwise subtract the smaller from the larger and use this, together with the smaller value, to repeat the calculation. The function shown in Fig. 27.3 evaluates this formula.

Fig. 27.3   Calculation of greatest common denominator

```
int gcd(int x, int y)     /*   gcd calculation   */
{
   if (x != y) {
      if (x < y)
         x = gcd(x, y-x);
      else
         x = gcd(y, x-y);
   }
   return x;
}
```

Enter this function and check that it works as expected. If further clarification is required then insert a `printf` function call to print the parameter values each time the function is called.

Recursion is a programming technique and so its usage is very subjective. Its use also requires some experience that can only really be gained by trial and error. Many recursive solutions can be re-written without using recursion. One possible rule for when recursion is required is that if the number of loops within a program is data–dependent then use recursion. Note that this is the number of loops, not the number of times a loop is repeated.

Summary

- Recursion is a programming technique used when a problem can be simplified in terms of itself.

Exercises

Unless already familiar with recursive methods an exercise along the lines 'write a program using recursion to...' is not realistic. Thus what follows is a project–style exercise leading through the implementation of a recursive function.

The problem is to solve the Towers of Hanoi game. The idea of this game is to move a series of different–sized rings from one peg to another via an intermediate peg as required. The rings start in an ordered pile with the largest at the bottom and the smallest at the top. Figure 27.4 shows the initial position where the pegs are labelled A, B, and C. The intention being to get all rings onto peg C.

Fig. 27.4 Towers of Hanoi starting position

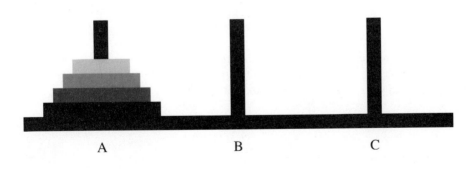

The rules for moving rings are:
(a) only one ring may be moved at a time.
(b) rings can only be placed either on empty pegs or on larger rings.

Before starting to write the program try a manual approach. Start with four coins of different sizes and try to find a solution. With four coins fifteen moves are required.

Solution

Now for the program. Initially there are three pegs (called A, B, and C) and there are four rings on peg A. The rings are to be moved to peg C. Moving four rings from A to C can be simplified to:

move three rings from A to B

move remaining ring from A to C

move three rings from B to C

Hence a recursive solution seems possible as moving four rings can be simplified to moving three rings twice plus a single move.

Another thing that falls out with the recursive approach is that when moving a stack of rings these are always smaller than any of the rings not being moved. Hence the rule of not putting a ring onto a smaller one is taken care of.

Start by writing a minimal program and in the main function call a 'move stack' function. This function should take four parameters as follows:

(a) type char — Source peg (' A')

(b) type char — Destination peg (' C')

(c) type char — Intermediate peg (' B')

(d) type int — Number of rings (4)

Next write this function. If the number of rings to move is one then move the ring. Do this using a function specifying the source and destination pegs as parameters. This function will be written shortly.

If the number of rings to move is not one, then use the move stack function to move one less than the number of rings from the source to the intermediate peg, move the remaining ring from the source to the destination, and then move the rings from the intermediate peg to the destination. You may need to re-read that description, possibly using some coins as an aid, before it makes sense.

Finally write the function to move the ring. All this need do is print a message saying 'Move ring from x to y' where x and y are the source and destination pegs. There is no need to specify any thing else.

28

Linked lists

Aims and objectives

The aims of this chapter are to:

- describe the idea of a linked list;
- show how they can be implemented using C;
- demonstrate how linked lists are used.

Overview

Linked lists are not a feature of the C language but a useful technique which can be achieved using C. Consider the problem of a firm processing employee records. Normally the output would probably be required in alphabetical form so this would be the most efficient way to store the data. However, suppose a program were required which listed employees by age. The data would be in the database alphabetically so the program would need to sort the data based on age. As each employee's record could contain a significant amount of data a sort method that required swapping records could be time–consuming. What would be preferred would be a method of specifying the order in which records are to be processed, or printed, rather than relying on the physical ordering in memory. This is exactly what linked lists can do.

Consider the structure defined in Fig. 28.1. It is designed to store the data about one employee.

Fig. 28.1 Structure for implementing linked lists

```
struct employee {
   char name[32];
   int age;
     .
     .
     .
   struct employee *next;
};
```

The key to linked lists is the last member. This is a pointer to a structure of type `employee`. Remember a structure cannot contain an undefined structure, but it can include a pointer to one, even to itself.

Back now to the problem of sorting the data. The sorting program either defines an array of type `struct employee`, or allocates structures as required, which does not matter. For ease of explanation the former will be used. Thus the program would start with an array of the form:

```
struct employee data[100];
```

Also one other variable would be required. This is a pointer and must be initialized to NULL. Typically it would be defined as:

```
struct employee *first = NULL;
```

This pointer is going to be the address in memory of the structure of the youngest employee: since no data has been read, it starts with a NULL value.

When the first employee record is read, it is saved in the first element of the array `data`. As this must be the youngest employee so far the pointer `first` is set to its address, that is,

```
first = &data[0];
```

Also the pointer within the structure (i.e. `data[0].next`) is set to NULL as there is no-one else. At this stage the list is as shown in Fig. 28.2.

Fig. 28.2   Initial linked list

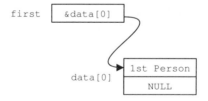

The second employee record is then read in. This one is saved in the next element in the array, `data[1]`. Let's assume that this person is older so needs to be printed after the previous one. To do this the following two assignments are required:

```
data[0].next = &data[1];
data[1].next = NULL;
```

Figure 28.3 shows the list after the addition of the second entry.

Fig. 28.3   Linked list with two items

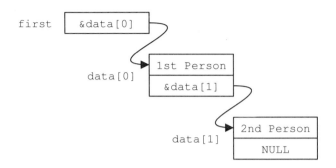

Finally (it is only a small firm) the last employee record is read. This person is aged between the other two. The record is saved in `data[2]`, but now we need to set the order by adjusting the pointers so that the `data[0]` points to this new entry, `data[2]`, which in turn points to the oldest employee, `data[1]`. This can be done as follows:

```
data[2].next = data[0].next;
data[0].next = &data[2];
```

Fig. 28.4   Complete linked list

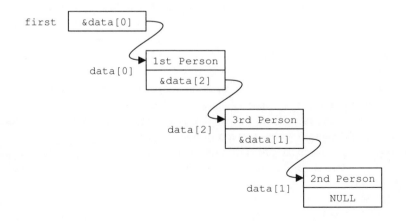

Figure 28.4 shows the completed linked list. The order of the data stored in memory is the same as read from disk but there is a different order which can be accessed using the pointers. A simple function to print out the employee names is shown in Fig. 28.5. The function requires the address of the first record to write, that is, the pointer variable `first`.

Fig. 28.5 Following a linked list

```
void prn(struct employee *ptr)
{
   while (ptr != NULL) {
      printf("%s\n", ptr->name);
      ptr = ptr->next;
   }
}
```

Creating a linked list

The code to do the printing is easier than that required to do the initial setting up of the pointers. The pointer definition will now be described using a simple word sort. The program is shown in Fig. 28.6.

The outer loop of the program reads words. As each word is read its correct position in the linked list is determined and the pointers changed to insert it. This process requires the use of a double pointer, that is, a pointer to another pointer. This is because the correct position in the list for the new word cannot be determined until the position has been passed. To put this another way: having reached a place in the list one can say 'the new word must be placed before this one'. In order to place before the current position, the address of the previous position must be kept, hence the use of the double pointer.

In the example shown in Fig. 28.6 the pointer variable `ptr` is used to keep the address of the pointer to the current position. Initially it is set to the address of the variable `first`. When scanning the list to determine the correct position the variable `p` is set to the current position. If this is NULL then the end of the list has been reached. If this is not NULL then the words are compared and the next item processed—this is known as traversing the list—until the correct position is found. At this point the variable `p` contains the address of the element to follow the new element, while `ptr` contains the address of the pointer which pointed to this element.

To add the new entry one saves the word, sets its next pointer to the value of `p` (the address of the element to follow this one), and sets the pointer addressed by `ptr` to the address of the new element.

This is quite a difficult concept to grasp, but it is worth going through the program by hand, working out the values of all the variables, possibly drawing diagrams, as per Figs. 28.2 – 28.4, to ensure correct understanding.

Fig. 28.6 Sorting words using linked lists

```c
#include <stdio.h>
#include <string.h>

int main(void)
{
   struct item {
     char word[16];
     struct item *next;
   };
   struct item data[50];
   struct item *first = NULL;
   struct item *p, **ptr;
   char word[16];
   int in = 0;

   /*  Loop for all data values  */

   while (scanf("%s", word) == 1) {
     ptr = &first;

   /*  Locate correct position in list  */

     while ((p = *ptr) != NULL) {
       if (strcmp(p->word, word) > 0)
         break;
       ptr = &(p->next);
     }

   /*  Add new entry in correct position  */

     strcpy(data[in].word, word);
     data[in].next = p;
     *ptr = &(data[in++]);
   }

   /*  Now print them out  */

   p = first;
   while (p != NULL) {
     printf("%s\n", p->word);
     p = p->next;
   }
   return 0;
}
```

Binary trees

The program shown in Fig. 28.6 demonstrates a technique, but is far from perfect. There is no check that the words do not overflow the available space and the sorting method is very simplistic. If a word occurs more than once then it is output more than once. A better word–sorting technique would be to build a binary tree. This requires each structure to have two pointers, one for words alphabetically before the current word, and one pointer for those following the current word. Such a structure is shown in Fig. 28.7. Note the two pointers and also a counter for the number of times the word has occurred.

Fig. 28.7   Binary tree structure definition

```
struct item {
   char word[16];
   int  count;
   struct item *left;
   struct item *right;
};
```

Adding the words in the form of a binary tree is much the same as shown previously, except that a decision has to be made to determine which pointer to follow. This is done by comparing the new word with the one in the structure (using strcmp) and taking one branch if the result is positive and the other if negative. Obviously a zero value indicates the word is already saved so just increment the count. Figure 28.8 shows a code fragment to locate the required position for the new word, and Fig. 28.9 shows the code to add the entry. The variable definitions are as for the program shown in Fig. 28.6. Note the extra test before adding the entry to allow for words already being in the list.

Fig. 28.8   Searching a binary tree

```
/*   Locate correct position in list   */

     ptr = &first;
     while ((p = *ptr) != NULL) {
        i = strcmp(p->word, word);
        if (i == 0) {
           p->count++;
           break;
        }
        if (i > 0)
           ptr = &(p->left);
        else
           ptr = &(p->right);
     }
```

Fig. 28.9   Adding a new entry

```
/*   Add new entry in correct position   */

    if (p == NULL) {                 /*   If new word   */
        strcpy(data[in].word, word);
        data[in].count = 1;
        data[in].left = NULL;
        data[in].right = NULL;
        *ptr = &(data[in++]);
    }
}
```

Once built the binary tree needs printing. This requires a modification to the earlier output and introduces a good use for recursion. Starting at the top of the tree one first prints all those words to the left of the current word. One then prints the current word followed by those to the right. Figure 28.10 shows a suitable function which is called as follows:

```
bprint(first);
```

Fig. 28.10   Printing a binary tree

```
/*   Function to print a binary tree   */

void bprint(struct item *p)
{
   if (p == NULL)
      return;
   bprint(p->left);
   printf("%3d - %s\n", p->count, p->word);
   bprint(p->right);
}
```

Summary

- Linked lists are a way of ordering items which is not dependent on the order these items are stored in memory.

- To implement a list involves using a structure containing a pointer. This pointer is used to create the list.

- Lists do not have to be limited to one pointer. Binary tree lists require two pointers per entry (one to the left and one to the right).

- It is not uncommon to have both forward and backward pointers. The forward pointers link to the next item in the list, while the backward pointers link to the previous item.

Exercises

1 Figures 28.7 to 28.10 show code fragments to implement a word sort using a binary tree. Combine these parts into a working program adding the required variable definitions, include files, and function prototypes. The code shown in Fig. 28.6 gives most of what is required.

2 Assuming the solution to exercise one followed closely the program shown in Fig. 28.6, there is a limit of the number of words permitted. This is determined by the size of an array. To avoid this limitation remove the array definition entirely and allocate (using `malloc`) a structure for each word when it is found not to be in the list. Do not forget to test the return value from `malloc`. This makes the program far more general. A word of caution: the recursive print function could exhaust the available stack space if a large number of words were entered. This is a problem with any recursive function.

29

DOS interface

Aims and objectives

The aims of this chapter are to:

- show how C programs can interface to both the BIOS and to DOS;
- describe the mechanism required;
- give some example usages.

Overview

On an IBM/PC—or clone—there are two useful pre-written sections of code to which one generally has access from within C. The first is the DOS operating system itself and the other is the BIOS. The BIOS is a set of routines which provide a standard interface between DOS and the machine hardware. Simple things like printing a character on the screen and reading a block from disk are handled by this software in a standard manner. Thus if new hardware is developed only the BIOS need change. The operating system can continue issuing the same requests to the BIOS and all should work. Generally this is a simplification as, if new features are added to the hardware, the operating system normally requires enhancing to make use of these extra features. Using BIOS or DOS functions rather than accessing the hardware directly will make programs more portable between various models and different clones.

It should be noted that the interface to specific hardware is not covered in the standard for C and thus what follows is specific to IBM/PCs.

Software interrupts

Requests to both DOS and the BIOS are done by invoking a software interrupt. The details of this can remain a mystery as generally a function is provided to do this. Which software interrupt is called depends on the function required. Values set in the processor's registers determine any sub–function if appropriate.

For example, the BIOS software interrupt 10 hex (16 decimal) controls the display. The value in register ah determines the sub function, a zero value requesting the screen mode to be set. In this case the screen mode is set to the value in al. Thus to select screen mode 3 (25 lines of 80 characters in 16 colours) then the following would need setting:

```
register ah = 0
register al = 3
execute software interrupt 10 hex
```

int86 function

This can be done using the int86 function. As this is not part of the ANSI definition of C, implementations other than those of Microsoft may use an alternative function name. The functionality should still be available. The int86 function requires three parameters. The first is an integer containing the software interrupt number. The last two parameters are both pointers to unions of type REGS. This is defined in the include file dos.h and follows much the same lines as described in Chapter 24.

The union specified by the second parameter defines the values of the registers required by the software interrupt while the last parameter contains the returned values. The example shown in Fig. 29.1 demonstrates a program to set the screen mode to three.

Fig. 29.1   Setting screen mode using the BIOS

```
#include <dos.h>
int main(void);

int main(void)
{
   union REGS rin, rout;
   rin.h.ah = 0;
   rin.h.al = 3;
   int86(0x10, &rin, &rout);
   return 0;
}
```

As with most standard functions int86 returns a value. This is the returned contents of the ax register (an unsigned int) and is often an indication of whether the function has worked or not. Alternatively, a failed request returns an indication in the sixteen–bit carry register (in the above example this would be "rout.x.cflag"). Generally a non-zero value indicates failure.

`intdos` function

Software interrupt 21 hex is the interface to DOS functions. It is so commonly used that a separate function is available with many compilers; for Microsoft C the function is called `intdos`. It requires two parameters that are pointers to two unions of type REGS. The following two calls are functionally the same:

```
int86(0x21, &rin, &rout);
intdos(&rin, &rout);
```

The DOS software interrupt offers many different functions and allows access to all the peripherals. A simple example is shown in Fig. 29.2 where function 2A hex is used to read the current date. Note how the same register union is used both to supply the values and for the returned values.

Fig. 29.2   Getting the date using the DOS interrupt

```
#include <stdio.h>
#include <dos.h>

int main(void);

int main(void)
{
    static char *days[] = { "Sunday",
                            "Monday",
                            "Tuesday",
                            "Wednesday",
                            "Thursday",
                            "Friday",
                            "Saturday" };
    union REGS regs;
    regs.h.ah = 0x2A;
    intdos(&regs, &regs);
    printf("Today is %s\n", days[regs.h.al]);
    printf("The date is %02d/%02d/%d\n", regs.h.dl,
            regs.h.dh, regs.x.cx);
    return 0;
}
```

Further reading

A full description of all the available BIOS and DOS calls is beyond the scope of this text. There are numerous volumes on this subject, one of the more readable being *Advanced MSDOS* by Ray Duncan (1986). Also for those with network access a list of most interrupts has been compiled by Ralf Brown and is available free from many bulletin boards. Any functions required for the projects will be described as needed.

Summary

- Although not part of the standard, most implementations of C for IBM/PCs supply a mechanism for accessing both the BIOS and DOS.

- Using these routines requires setting the required values in a suitable data structure and then executing a software interrupt.

Exercises

1   The BIOS software interrupt 10 hex is the general display control mechanism. Various sub–functions allow screen output to be controlled; see Appendix E for examples. Sub–function fifteen (0F hex) determines the current screen mode. The current mode is returned in register `al`. Write a short program along the lines of that shown in Fig. 29.1 which uses interrupt 10 hex, sub–function 0F hex to determine, and then print out, the current screen mode. Verify that the result is reasonable by comparison with the values given in Appendix C, or by using your solution to the first exercise in Chapter 16. Note that the BIOS mechanism is preferred to direct memory reference as it is more portable, although there is a slight increase in processor time used.

2   DOS function 2C hex is used to return the current time. The values are returned in registers as follows: `ch` = hours, `cl` = minutes, `dh` = seconds. Modify the program shown in Fig. 29.2 to print out the time of day using the twenty four hour clock (i.e. as returned by the DOS call).

3   Now modify your previous solution to use the twelve hour clock, correctly inserting the required a.m. and p.m. suffixes.

30

Project introduction

Well, it's easy if you understand it.

J. M. Jarvis (1991)

Aims

The aims of this chapter are to:

- describe a structured approach to program design;
- give some hints on approaching the projects;
- detail how to get the project data.

Introduction

The projects are included in the text because no language can be learned without suitable practical sessions. The projects are designed to be interesting but clearly not all will appeal to everyone. Before embarking on the projects, a few words on program design.

Structured programming

In Chapter 8 it was stated that C was a language that could support a structured approach, but no attempt was made to define such an approach. As mentioned earlier, structured programming is a vast topic and all that will be described here is just a brief summary.

There are two general approaches to program design. These are called 'top down' and 'bottom up'. Both will now be summarized, but it is the top down approach that the author recommends. In real life often a mixture of the two methods is used—probably with a bit of experimentation as well.

Top down design

The global idea with this method is first to define the problem in general and then to refine the definition until all the detail is complete. Defining the intention of a program is often missed! Start by stating the obvious: 'this program is designed to ...'. From here define how the program will achieve this end, what functions are required, what data is needed, and—possibly most importantly—what assumptions are made.

Start by writing the main program. Where functions are required initially just define stubs. These are functions which do nothing other than indicate their presence. For example, an input function might be required. The initial stub for this function would simply return either some test data, or an error condition. This would enable the main program to be tested without all the required functions being completed. Clearly the program would not be complete until all the stubs have been converted to fully working functions, but it is surprising how much can still be tested.

Once happy that the main function is working, take each of the stubs and expand them to fully working functions. This may require creating more stubs which are used as before. The program development progressively refines the code, creating stubs where the full detail is not yet required, and converting stubs to fully working functions.

A useful side effect of this method is that—providing the functionality of a stub is fully defined—another programmer can fill in the details and complete it.

Bottom up design

This approach is the inverse of the above. One starts by writing the detailed functions first and then combining them until one reaches the complete solution.

This method often re-uses existing functions and knowledge and is often used in the real world; however, it is suggested that this technique is reserved until one has gained more experience with programming in C.

Projects

Follow the suggestions given with each project. The text shows a suggested path to the solution, using the top down approach, and includes breaks for testing and assimilating progress. Do not be tempted to rush into a total solution. It is not uncommon for programmers to write an entire program of a few thousand lines and spend days debugging it, only to find that although the complicated transformations and graphical output was fine, it was the data that was not being read correctly.

Each project lists at the start the main features required to solve it. But beware, there is no unique solution so it may be that your answer might not use all the expected features. This does not necessarily mean it is wrong; it could just be better.

The question of how many functions a specific program should contain is very difficult to determine. Generally a program can be split into an input, a processing, and an output phase, possible contenders for three separate functions. As another general rule, restrict functions to about fifty lines, two screens full.

Solutions

As you will no doubt have already discovered, solutions are supplied. These are not designed to be read! It would be nice not to have to refer to these at all, but it is expected that some guidance may be required. It is for this reason that the solutions appear. As stated earlier, reading another's solution is very much easier than developing one's own, and teaches one far less.

The solutions to the various stages of the projects are supplied in the order suggested. There is then a complete solution at the end. Whenever possible, refer only to the parts needed. This will avoid accidentally seeing the next section!

Project data

Some of the projects require data files. Generally these can be entered manually but there are two projects where this is not so. Both the TIFF data file and the satellite data files have to be obtained either by downloading from the anonymous ftp site `src.doc.ic.ac.uk`, or by obtaining a data disk by completing the form at the back of this book. Using the ftp site is the preferred route as this enables larger image files to be acquired. The file names and their contents are summarized in the file `readme.txt`, both in the ftp directory and on the disk. This file should be consulted to establish the contents of the remaining files.

Anonymous ftp

To get the data files via ftp proceed as follows:

- `ftp src.doc.ic.ac.uk`
- Enter `anonymous` as the user name
- Enter your mail name as the password
- Change directory to `packages/Intro-to-C`
- Retrieve the summary file (`readme.doc`) and the required data files. Do not forget to use binary transfer mode for the TIFF and image files.

31

Simple numeric sorting

Uses

The solution will probably require the following:

- passing arguments to the main function;
- passing arrays to functions;
- file input;
- use of flag variables.

Overview

Sorting numbers is not the most exciting of projects but, being the first, it does keep things rather simple. The idea is to read a set of integer numbers from a specified file, and then print them to the screen in ascending order. One of the first items required will be a suitable data file so it is worth creating one with any suitable editor. Numbers should be integer and there should be only one per line.

Getting started

First write a main function that requires a filename as a parameter. The program should check that a filename has been supplied; if not, write some suitable message and exit. If a file name has been specified, then the file should be opened in read mode. Check the return value from `fopen` and again issue an error message if there is a problem. Once the file has been opened print a suitable message, close the file, and exit. Remember a zero exit value indicates all went well, a non-zero value indicates an error. This much should be tested before progressing.

If you have difficulty with this, refer back to Chapter 22 and particularly to your answer to the first exercise. Make sure you understand this code as it forms the basis of many programs.

Reading the data

Once the first part is working, add a loop to read each value from the file and simply print the value. While `fscanf` could be used here, it is not recommended because of the problem with illegal input. Once an illegal input character is reached, `fscanf` will not pass this character. A better approach would be to use `fgets` to read the line, and then `sscanf` to convert the line to a numeric value. Two separate checks are required, one to check that `fgets` has not reached the end of file, and one to check that `sscanf` correctly translates the number.

A prudent programmer would also check that each input line was not longer than the character array allocated. This could be done by looking for the '\n' character in each line read. The standard function `strchr` (mentioned in Chapter 26, Table 26.2) returns the address of the first occurrence of a given character in a character string. A NULL return value indicates that the character was not found. If this situation arises it is probably a good idea to abort the program.

Again, test this much before progressing. What happens if the file exists but there are no data in it? Is this reasonable? Check also the result of having illegal characters in the data and also lines that are too long.

Saving the data

Having read and printed the data, modify the program to store the data into an array of type `int`. This array should have plenty of space, say 100 elements. As the data are read keep a count of the number of items entered as this will be required later.

Earlier it was stated that creating a fixed size array to process an unknown amount of data can be very wasteful. For this example let's not worry about this, just define a suitably large array within the main function but beware of creating too large a dynamic array. As explained earlier, large dynamic arrays may cause the stack to overflow.

Sorting

Next add a function called `sortint` which is passed as parameters, the array containing the data, and the number of items in that array. Do not forget to define a function prototype. Within this function set up a loop to print all the data values. This will check that the data have both been read correctly and been passed into the function properly.

Once happy that the data have been read in, it is time to sort them. There are many different sorting algorithms, the pros and cons of which will not be dealt with here. One of the easier techniques is the bubble sort. While this is a very inefficient algorithim it is straightforward to implement. Readers interested in sorting algorithms would do well to look at *The art of computer programming, Vol 3 — Sorting and Searching* by D. Knuth (1973).

The bubble sort involves scanning the data, and swapping elements that are not in the correct order. The data are rescanned until no swap is required; the data are then in order. Thus the sorting function requires two parts; a loop which is repeated until the data are in order and a second loop which scans every element of the data, reversing values as required.

Start with the second bit first. Write a loop that compares all adjacent elements in the data array and swaps those not in the correct order. Define an extra variable of type `int` that is set to zero (false) each time two numbers are swapped.

The last bit of the function requires the outer loop to keep repeating the sort loop until the data are in order. The data must always be scanned once so this looks like a good place to use a `do-while` loop. The condition that this loop should be repeated is that at least one data swap occurred in the inner loop. The flag variable being set to zero will indicate this. Thus before the inner sort loop is executed, but within the outer loop, this flag variable should be set to a non-zero (true) value. So the structure of the sorting function should be something like this:

- Start a `do-while` loop
- Set a flag variable to true
- Scan all data values and, if a swap is required, do the swap and
  set the flag variable to false
- Repeat `do-while` loop while the flag variable is false

Finally, print the contents of the data array after it has been sorted. This should probably be done in the main function leaving the sort function to do just the sorting process. It is always a good idea to limit the processing in a function to one specific task. It makes the final program more readable and easier to change. For example, if a different sorting method were to be used just the one function would need to be changed.

Sorting project—getting started

```
/*   Numerical data sorting program */
/*   P. Jarvis.             09/02/1991 */

#include <stdio.h>

int main(int, char **);
void exit(int);

int main(int argc, char *argv[])
{
   FILE *fh;

/*   Check filename specified (i.e. two arguments)   */

   if (argc != 2) {
      fprintf(stderr, "Usage:   %s filename\n", argv[0]);
      exit(1);
   }

/*   Try to open the specified file   */

   fh = fopen(argv[1], "r");
   if (fh == NULL) {
      fprintf(stderr, "Unable to open %s for input\n", argv[1]);
      exit(2);
   }

/*   All is well so say so and clear up   */

   printf("Ready\n");
   fclose(fh);
   return 0;
}
```

Sorting project—reading the data file

```
/*   Numerical data sorting program */
/*   P. Jarvis.           09/02/1991 */

#include <stdio.h>
#include <string.h>

#define MAXNUMS 100

int main(int, char **);
void exit(int);
void sortint(int *, int);

int main(int argc, char *argv[])
{
  FILE *fh;
  char line[80];
  int i, num;
  int data[MAXNUMS];

/*   Check filename specified (i.e. two arguments)   */

  if (argc != 2) {
    fprintf(stderr, "Usage:   %s filename\n", argv[0]);
    exit(1);
  }

/*   Try to open the specified file   */

  fh = fopen(argv[1], "r");
  if (fh == NULL) {
    fprintf(stderr, "Unable to open %s for input\n", argv[1]);
    exit(2);
  }

/*   Read data until either MAXNUMS numbers have been read   */
/*   or the end of the file is reached.                      */

  num = 0;
  while ((num < MAXNUMS) && (fgets(line, 80, fh) != NULL)) {
    if (strchr(line, '\n') == NULL) { /*  No '\n' so error */
      fprintf(stderr, "Error - line longer than 79 chars\n");
      continue;
    }
    if (sscanf(line, "%d", &data[num]) != 1)
      fprintf(stderr, "Illegal input - %s (ignored)\n", line);
    else
      num++;
  }

/*   Next print out the numbers read   */

  for (i=0; i<num; i++)
    printf("%d\n", data[i]);

/*   Finally clear everything up   */

  fclose(fh);
  return 0;
}
```

Sorting project—sort function

```
void sortint(int *data, int num)    /*  Sort integer data   */
{
  int i, j, sorted;
  do {
     sorted = TRUE;                  /*  Assume in correct order  */
     for (i=1; i<num; i++) {         /*  For every number         */
        if (data[i] < data[i-1]) {   /*  Swap if out of order     */
           j = data[i];
           data[i] = data[i-1];
           data[i-1] = j;
           sorted = FALSE;           /*  Flag not yet sorted      */
        }
     }
  } while (!sorted);                 /*  Repeat until sorted      */
}
```

Sorting project—complete solution

```
/*   Numerical data sorting program */
/*   P. Jarvis.            09/02/1991 */

#include <stdio.h>
#include <string.h>

#define TRUE 1
#define FALSE 0
#define MAXNUMS 100

int main(int, char **);
void exit(int);
void sortint(int *, int);

int main(int argc, char *argv[])
{
   FILE *fh;
   char line[80];
   int data[MAXNUMS];
   int i, num;

/*   Check filename specified (i.e. two arguments)   */

   if (argc != 2) {
      fprintf(stderr, "Usage:   %s filename\n", argv[0]);
      exit(1);
   }

/*   Try to open the specified file   */

   fh = fopen(argv[1], "r");
   if (fh == NULL) {
      fprintf(stderr, "Unable to open %s for input\n", argv[1]);
      exit(2);
   }

/*   Read data until either MAXNUMS numbers have been read   */
/*   or the end of the file is reached.                      */

   num = 0;
   while ((num < MAXNUMS) && (fgets(line, 80, fh) != NULL)) {
      if (strchr(line, '\n') == NULL) { /*   No '\n' so error
*/
         fprintf(stderr, "Error - line longer than 79 chars\n");
         continue;
      }
      if (sscanf(line, "%d", &data[num]) != 1)
         fprintf(stderr, "Illegal input - %s (ignored)\n", line);
      else
         num++;
   }

/*   Now sort the data   */

   sortint(data, num);

/*   Then print out the result   */

   for (i=0; i<num; i++)
      printf("%d\n", data[i]);

   fclose(fh);
   return 0;
}
```

```
void sortint(int *data, int num)    /*   Sort integer data   */
{
    int i, j, sorted;
    do {
        sorted = TRUE;                  /*   Assume sorted            */
        for (i=1; i<num; i++) {         /*   For each number          */
            if (data[i] < data[i-1]) {  /*   Swap if required         */
                j = data[i];
                data[i] = data[i-1];
                data[i-1] = j;
                sorted = FALSE;         /*   Flag not in order        */
            }
        }
    } while (!sorted);                  /*   Repeat until sorted      */
}
```

32

File analysis

Uses

The solution to this project will probably require:

- passing arguments to the main function;
- processing a variable number of parameters;
- using variables as format descriptors;
- octal, decimal, and hexadecimal output

Overview

The idea of this project is to write the contents of a given file as a sequence of bytes. These bytes are printed in octal (base 8), decimal (base 10), or hexadecimal (base 16) format, depending on a parameter. This process—known as dumping a file—enables the exact contents of a file to be examined. A working solution to this problem may help with later projects. It is suggested that the output format is along the following lines:

```
000000    043 061 070 034 015 012 062 072
000010    033 034 036 037 000
```

The above output assumes an octal format has been requested. The six–digit number at the start of each line is the file offset (i.e. position within the file) where the following data occur. This number uses the same number base as the rest of the line. There then follows the actual bytes from the file each separated by a space, and written in the requested base. The number of bytes written on each line should be the same as the numeric base used to output the data. Compare the first example, which uses an octal format, with the following, which uses a decimal format, and check that you understand what is required.

```
000000    35  49  56  28  13  10  50  58  27  28
000010    30  31   0
```

The next design decision to be made is how to specify the required numeric base to use. One could use the word 'octal', if one required a base–eight analysis, but how would one then print the contents of a file called octal? It is suggested that the Unix approach is adopted. To distinguish between the options and the file names, the options are prefixed with a minus sign. Thus to print the contents of the file 'fred' in octal one would use:

```
dump -o fred
```

Note how the numeric base has been specified using only a single character. This is not mandatory, but probably simplifies things a bit. It would also be nice to permit more than one filename to be specified so the following should also be valid:

```
dump -o fred bill harry
```

Getting started

Initially write a program that takes a single filename as a parameter. This file should be opened and then closed. Suitable error messages should be printed if the filename is not specified, or if the file cannot be opened. Remember a zero exit value indicates all went well.

Next modify the above so that the file is opened, in binary mode, within a function. The function should take one parameter, the file name as a character string. Change the main function to call this and repeat the testing as above. The function should return a false (zero) value if an error occurred, and a true (non-zero) value otherwise. The return value should be checked in the main function and a suitable exit made.

Astute readers may have noticed a slight contradiction in the return values specified in the previous two paragraphs. The function main should return a zero value if it worked, whereas the dump function returns a zero (i.e. false) value if it fails.

The return value from main is defined, zero means the program worked. The return value from any user written functions is entirely up to the programmer. By convention the author tends to use a non-zero return value (i.e. true) if the function behaved correctly, and zero otherwise. This of course assumes that only a pass or fail indication is required.

Dumping in a fixed base

Next the function needs to be extended to print the contents of the file. This will require a counter to keep track of the byte position within the file (the file offset) and another to keep the current byte. Both could be of type int, although the first might require long for larger files. Start with dumping the file in octal. A new line of output will be required each time the file offset is a multiple of eight, in other words, when offset%8 is zero. Thus a loop with the following form will be required.

- Read a character from the file.
- If it is not the end of file, then process as follows.

    If the file offset modulo eight is zero, output a new line and the offset as a six–digit octal number ("%06o" format string). Remember to use the form "%06lo" if the offset is of type long.

- Finally print the byte read, as octal, " %03o" would be an appropriate format string.

Experiment using a small file, possibly the data file for the sorting project, and adjust the printf statements for the most pleasing effect. A new line character may be required at the end of the dump. Do not forget to return a true value from the dump function on successful completion.

Preparing for a variable base

At this stage highlight each part of the program which would need modification if the base were changed from octal to say decimal. There should be only three such places. Add a local variable to the function that is going to indicate which base the dump is to be in. Valid values for this variable will be 8, 10, and 16. Preset the value to eight. Next define two character string pointers, presetting them as follows:

```
static char *ptr1 = "%06o   ";
static char *ptr2 = " %03o";
```

Remember to use "%06lo " if the offset is of type long. Modify the function to make use of the three variables just defined and one should find that all the dependencies on the base are now controlled by these variables.

Dumping using a variable base

Next the value for the base needs to be passed as a parameter. Add this, again using a default of octal from the calling function. Check that this much works.

Then add a switch statement at the start of the dump function to set the character string pointers depending on the value of the requested base. Remember that the only valid values will be 8, 10, and 16. Safety conscious programmers will, of course, allow for other values. Check that this works with each of the three valid bases, and an invalid one, if this has been allowed for.

All that now remains is to allow the required base to be specified as an argument to the program. Extend the parameter processing to include a loop so that every given argument is processed. If an argument begins with a minus sign then check the next character and set a variable to 8 if the letter is o, 10 if d, and 16 if x. Allow for both upper and lower case. A switch statement is probably the best method here. If the argument does not begin with a minus sign then it is assumed to be a filename and the dump function called using the latest value of the base variable. Note that the base will require a default value. Make these changes and check that all is well. Try commands of the following form and see if the program behaves as expected:

```
dump -o filename
dump -o filename -d filename
dump -d filename filename
```

Paging the output

One final modification. It would be nice if the dump program paused after every twenty or so lines and did not continue output until a key was pressed. This would enable the output to be read before it scrolled off the top of the screen. To do this, define a variable in the dump function and initialize it to a value of twenty. Each time a new line of output is started, decrement this counter and, if zero, wait for a character to be entered. If the `getchar` function is used then the Enter key will be required to continue but if the `getch` function is used any character will do. The `getch` function requires the include file `conio.h` and is not part of the ANSI standard but is included in most DOS implementations. It enables unbuffered input from the keyboard.

File dump project—getting started

```
/*     File dump utility     */
/*   P. Jarvis.    23/03/91   */

#include <stdio.h>

int main(int, char **);
void exit(int);

int main(int argc, char *argv[])
{
   FILE *fh;
   if (argc != 2) {
     fprintf(stderr, "Usage:   %s   filename\n", argv[0]);
     exit(1);
   }
   fh = fopen(argv[1], "rb");
   if (fh == NULL) {
     fprintf(stderr, "Unable to open %s\n", argv[1]);
     exit(2);
   }
   fclose(fh);
   return 0;
}
```

File dump project—initializing the dump function

```
/*      File dump utility     */
/*   P. Jarvis.    23/03/91   */

#include <stdio.h>

#define TRUE 1
#define FALSE 0

int main(int, char **);
void exit(int);
int dump(char *);

int main(int argc, char *argv[])
{
   if (argc != 2) {
      fprintf(stderr, "Usage:   %s   filename\n", argv[0]);
      exit(1);
   }
   if (!dump(argv[1])) {
      fprintf(stderr, "Dump function failed\n");
      exit(2);
   }
   return 0;
}

int dump(char *filename)
{
   FILE *fh;
   fh = fopen(filename, "rb");
   if (fh == NULL)
      return(FALSE);
   fclose(fh);
   return TRUE;
}
```

File dump project—dumping in octal only

```c
/*      File dump utility    */
/*   P. Jarvis.   23/03/91   */

#include <stdio.h>

#define TRUE 1
#define FALSE 0

int main(int, char **);
void exit(int);
int dump(char *);

int main(int argc, char *argv[])
{
   if (argc != 2) {
      fprintf(stderr, "Usage:  %s  filename\n", argv[0]);
      exit(1);
   }
   if (!dump(argv[1])) {
      fprintf(stderr, "Dump function failed\n");
      exit(2);
   }
   return 0;
}

int dump(char *filename)
{
   FILE *fh;
   long offset = 0L;
   int c;

   fh = fopen(filename, "rb");
   if (fh == NULL)
      return(FALSE);

   while ((c=fgetc(fh)) != EOF) {
      if ((offset%8L) == 0L)
         printf("\n%06lo ", offset);
      printf(" %03o", c);
      offset++;
    }
   printf("\n");
   fclose(fh);
   return TRUE;
}
```

File dump project—processing parameters

```
/*      File dump utility    */
/*   P. Jarvis.    23/03/91   */

#include <stdio.h>

#define TRUE 1
#define FALSE 0

int main(int, char **);
int dump(char *, int);

int main(int argc, char *argv[])
{
   int i;
   int base = 8;
   for (i=1; i<argc; i++) {
      if (argv[i][0] == '-') {
         switch (argv[i][1]) {
            case 'd':
            case 'D':
                     base = 10;
                     break;
            case 'h':
            case 'H':
                     base = 16;
                     break;
            case 'o':
            case 'O':
                     base = 8;
                     break;
            default:
                     fprintf(stderr, "Illegal parameter - %c\n",
                                  argv[i][1]);
                     break;
         }
      }
      else {
         if (!dump(argv[i], base))
            fprintf(stderr, "Dump failed for - %s\n", argv[i]);
      }
   }
   return 0;
}
```

File dump project—complete solution

```
/*      File dump utility    */
/*   P. Jarvis.   23/03/91   */

#include <stdio.h>
#include <conio.h>

#define TRUE 1
#define FALSE 0

int main(int, char **);
void exit(int);
int dump(char *, int);

int main(int argc, char *argv[])
{
   int i;
   int base = 8;
   for (i=1; i<argc; i++) {          /*   for every parameter   */
      if (argv[i][0] == '-') {
         switch (argv[i][1]) {       /*   Validate specified base   */
            case 'd':
            case 'D':
                     base = 10;      /*   Decimal   */
                     break;
            case 'h':
            case 'H':
                     base = 16;      /*   Hexadecimal   */
                     break;
            case 'o':
            case 'O':
                     base = 8;       /*   Octal   */
                     break;
            default:
                     fprintf(stderr, "Illegal parameter - %c\n",
                              argv[i][1]);
                     break;
         }
      }
      else {
         if (!dump(argv[i], base))
            fprintf(stderr, "Dump failed for - %s\n", argv[i]);
      }
   }
   return 0;
}
```

```c
/*   Dump specified file in given base   */

int dump(char *filename, int base)
{
   FILE *fh;
   long offset = 0L;
   int line = 20;
   char *s1, *s2;
   int c;

   switch (base) {                      /*   Check given base   */
      case 8:
              s1 = "\n%06lo ";
              s2 = " %03d";
              break;
      case 10:
              s1 = "\n%06ld ";
              s2 = " %03d";
              break;
      case 16:
              s1 = "\n%06lx ";
              s2 = " %02X";
              break;
      default:
              return FALSE;
   }

   fh = fopen(filename, "rb");       /*   Open the file   */
   if (fh == NULL)
      return FALSE;

   while ((c=fgetc(fh)) != EOF) {
      if ((offset%(long)base) == 0L) {
         if (line-- == 0) {                        /*   End of page?   */
            getch();
            line = 20;
         }
         printf(s1, offset);                       /*   Print offset   */
      }
      printf(s2, c);                               /*   Print byte   */
      offset++;
   }
   printf("\n");
   fclose(fh);
   return TRUE;
}
```

33

Generating PostScript

Uses:

The solution to this project will involve:

* passing arguments to the main function;
* simple character processing.

Overview

PostScript, like C, is a programming language. However, unlike C, PostScript is specifically targeted to producing text and graphics on a printed page. It is a language which defines character position, size, shape, and colour. How to draw lines and pictures is also specified. It is not the intention to teach another complete language: only a few basic commands will be mentioned.

The aim of this project is to write a standard text file to a laser printer using the PostScript page description language. In order to do this, the file will need to be converted from plain text into the PostScript language. Once written the program will then be able to drive any PostScript–compatible printer, including high–resolution typesetters.

PostScript commands

There are only three commands you will need to know. These are `moveto`, `show`, and `showpage`. There are a few other commands required but these are supplied for you in a function which is described later.

`moveto` defines the current point. This is the position on the page where any subsequent text will be placed. Because of the nature of the language the integer x and y co-ordinates precede the command, thus a typical example would be `20 255 moveto`. The co-ordinate system has its origin at the lower left corner of the page and, by default, works in points (1 point = 1/72 inches).

`show` writes the given text at the current position on the page. The text must precede the command and also must be enclosed in parentheses. If the text contains either type of parentheses then these must be escaped by prefixing with a backslash. Similarly two backslash characters are required to represent a single backslash. In other words: a \ becomes \\, a (becomes

\ (, and a) becomes \). Other than these three special cases, all characters are passed through without change. Thus to write the text 'Hello, it's me (Paul)' at position (20, 255) would require:

```
20 255 moveto
(Hello, it's me \(Paul\)) show
```

The last command, `showpage`, actually puts ink onto the paper. Until this command is called, text and graphics are only stored in the printer's memory. When called, one page is printed.

Getting started

As with all projects, first start by writing a main function. This should accept a single filename as a parameter. If this is omitted then either an error can be flagged or, following the Unix convention, the input can be read from standard input, the latter being the more useful.

If a filename is specified then the given file should be opened in read mode. Do not forget to check the return value from `fopen` for an error condition. If no file name has been specified (indicated by `argc` being one) then the file handle should be set to the predefined handle `stdin`.

Finally in the main function add a loop which reads one character at a time from the input file until the end of input is reached. Each character read should initially just be written to the screen. On reaching the end of data close the source file and exit the function. Do not forget to return a meaningful exit value.

Enter this much and check that it works. Ensure that if no filename is specified then the input is correctly read from the keyboard. Check for correct behaviour if the specified file does not exist.

Character processing

Next add a function that is passed each character as it is read from the input file. The screen output in the main function will no longer be needed and should be removed. The character–processing function will put each character passed into a line buffer (i.e. a character array) and call a line–processing function when either the line buffer is full, or a new line character (\n) is read.

One other special character that will require processing differently by this function is the formfeed character (\f), which causes following output to begin on a new page. Implementation of this feature will be left until later. For the moment just detect the character and ignore it.

To summarize, any character other than a newline or form feed character will be placed into a line buffer. If this becomes full then a line–processing function will be called. The line processing function will also be called if a newline character is read and, possibly, if a formfeed character occurs.

Ignoring the special cases, start by adding each character to a line buffer and, if it becomes full, use a printf statement to write it out. This will require creation of a character array to hold the line and an integer variable to keep a count of the number of characters currently stored. Both these variables will need to be of type static otherwise their values will be lost from one call to the next. A suitable length for the line buffer would be 81 characters, this being the width of a normal text screen plus one to hold the end of string delimiter. Before a character is added to the line buffer check that there is room. If not, then add the string–terminating character (\0) at the end of the line and write the line out before inserting the new character. Check that this much works. You may find that the last few characters of a file are not printed. This is because they are still in the line buffer and this will have to be sorted out later.

In the case of extreme hardship, the code shown in Fig. 33.1 would be a good start for the processing described above.

Fig. 33.1   Adding characters to a line buffer

```
if (in == 80) {
   line[in] = '\0';
   printf("%s", line);
   in = 0;
}
line[in++] = (char) c;
```

Now add a test before the previous code to check for a newline character (\n). As the formfeed character will be processed later it would be worth using a switch statement at this point. The saving of a character into the line buffer would form the default case. The code required to process the newline character will save an end of string delimiter, but not the newline character, in the line buffer and then call the line–processing routine. Do not forget to reset the number of saved characters back to zero after calling the line–processing function. Add this code temporarily using a printf statement as the line processor. If the format string used is "%s\n" then the output should be identical to that of the previous test. If all is well go to the line–processing routine.

Line processing

This function has a number of requirements. For most lines it will output, to standard output, the required PostScript commands to move to the correct position on the paper and then write the text. Special processing will be required when a page becomes full, and if the text string includes characters that need prefixing. Before being able to determine the required position for a line on the page, the co-ordinate system and units to be used must be understood.

As mentioned earlier, the units used by default are points (i.e. 1/72 inch). The page is to be printed double columned and in landscape mode, the longest edges of the paper being the top and the bottom. The code to set this up will be supplied, but five values need to be available. These are listed in Fig. 33.2 and it is suggested that they are created as symbols using the `#define` pre-processor directive.

Fig 33.2   Suggested page layout symbols

```
TOP_MARGIN           536
BOTTOM_MARGIN         24
LEFT_MARGIN           30
RIGHT_MARGIN         424
LINE_HEIGHT            9
```

The top and bottom margins specify the vertical limits for text lines while the left and right margins give the left–hand position of the start of text in the left and right columns. The line height is the distance to move down for each line. Thus the initial co-ordinate for the first line in the left column would be (30, 536), while the second line would start at (30, 527).

Add two global variables to store the current text co-ordinates as integer values. They should be preset to the upper left corner initially but this will be changed later. The first thing that the line processing function should do is check these values. If the y co-ordinate is below the bottom margin then a new column is required. If it is the bottom of the left column then reset the co-ordinates to the top of the right–hand column. If the bottom of the right–hand column has been reached then reset to the top of the left column but make a note that something will be required here to start a new page.

Having processed the co-ordinates as above then write them out and create a PostScript `moveto` command that, for the first line, would be:

```
30 536 moveto
```

Next the text of the line needs processing. As mentioned earlier, text must be enclosed in parentheses and the three special characters, (,), and \ each need to be prefixed with a \. The line ends with the PostScript command show.

Thus start by writing a left parenthesis, then check each character of the given line. If it is one of the three special characters, output the prefix character \. Then output the character. When the end of the string has been reached output) show and a newline character to complete the command. Intermixing printf and fputc is perfectly valid. The final requirement for this function is to update the current *y* co-ordinate by subtracting the line height value defined earlier.

Check this much on a file that contains sufficient lines to ensure at least three columns of output will be produced. As before the last few characters may not be printed, but the important thing to check for here is that the co-ordinate system is correctly updated between each line, and from one column to the next.

Handling pages

Figure 33.3 lists a suitable newpage function. This function must be called before each page and draws a nice box with a title and page number. The function requires the title and page number to be passed as parameters, and requires the page number to start from one. The function rotates the printing axis so the output comes out in landscape mode and double columned.

In the line–processing function add a call to this function where it was noted that it would be required. Force a page number of one and some string constant for the title line. If you look closely at the conditions under which the newpage function will be called you will see that it is called when the column changes from right to left. To ensure the very first page is treated correctly, change the preset values for the *x* and *y* co-ordinates to be a point below the bottom of the right–hand column. Now re–run the program and confirm that the new page PostScript commands are added at the beginning and also each time the column changes from right to left. Check the co-ordinates are still processed correctly.

Two more variables are required at this stage. One is an integer page number and the other is a character string for the title. The page number variable should be static and defined within the line–processing function. Each time the function newpage is called the page number should be incremented. As explained earlier, the initial value must be equal to one.

Fig. 33.3 PostScript new page function

```
void newpage(char *title, int page)
{
   if (page == 1)
      printf("%%!PS-Adobe\n");
   else
      printf("showpage\n");
   printf("90 rotate);
   printf("0 -595 translate\n");
   printf("20 20 moveto\n");
   printf("816 20 lineto\n");
   printf("816 550 lineto\n");
   printf("20 550 lineto\n");
   printf("closepath stroke\n");
   printf("418 20 moveto\n");
   printf("418 550 lineto\n");
   printf("stroke\n");
   printf("/Times-Roman findfont\n");
   printf("20 scalefont setfont\n");
   printf("20 555 moveto (%s) show\n", title);
   printf("730 555 moveto (Page %d) show\n",
          page);
   printf("/Courier findfont\n");
   printf("7.9 scalefont setfont\n");
}
```

The title character string variable should probably be a global variable. It will need defining within the main function, either as the name of the input file (if specified this will be argv[1]), or as a predefined string (e.g. standard input). Verify that these changes work as expected.

Flushing the last lines

Each page is printed only when the PostScript showpage command is executed. This is automatically output at the start of each new page, other than the first, by the supplied newpage function. But what about the last partial page? When the end of input data has been reached then any partial line and any partial page must be printed.

To do this modify the process character function so that it is passed an end of file (EOF) character the following action is taken. If the current line is not empty, terminate the line and call the line processing function. Whether or not the line was empty, if the current text co-ordinates are anything other then the top left of the page, write out a PostScript showpage command.

The main function will also need a minor modification to ensure that the end of file condition is passed to the process character function. This is probably a good case for a `do` loop.

Check that after making these modifications that the final line of output is the command `showpage`. To be thorough a test file should be created that will exactly fill one printed page, both columns. When this is processed there should be only one `showpage` command at the end, two would create a blank page.

Processing formfeeds

The formfeed code (`\f`) should cause a skip to the top of the next column. To add this feature modify the process character function. Earlier a `switch` statement was suggested to allow for this extension. If not already done, add another case to process the formfeed character which is treated as follows. If the line buffer is not empty then it should be terminated and the line processing function called. Then by setting the current *y* co-ordinate to below the bottom margin a new page will be forced for any further text written.

Extensions

If you have access to a PostScript printer try the program for real. It might be nice to include a date and time in the title, for example. How to get the date and time from the DOS operating system was covered in Chapter 29. What happens if a number of blank lines are part of the input data? Could this be made more efficient?

Also how should tab characters (`\t`) be processed? It would be possible to replace tabs by say five spaces, or to set tab stops and output the required number of spaces to reach the next stop. This is left as an extension for those interested in actually using this program in production.

PostScript project—getting started

```c
/*   Text to PostScript  converter     */
/*   P.  Jarvis              23/03/1991   */

#include <stdio.h>

int main(int, char **);
void exit(int);

int main(int argc, char *argv[])
{
   FILE *fh;
   int c;

   switch (argc) {
     case 1:
             fh = stdin;
             break;
     case 2:
             fh = fopen(argv[1], "r");
             if (fh == NULL) {
                fprintf(stderr, "Unable to open %s\n",
                             argv[1]);
                exit(1);
             }
             break;
     default:
             fprintf(stderr, "Usage:  %s [filename]\n",
                             argv[0]);
             exit(2);
   }

   while ((c=fgetc(fh)) != EOF)
      putchar(c);

   fclose(fh);
   return 0;
}
```

PostScript project—initial process character function

```c
void proc_char(int c)          /*   Process character   */
{
   static char line[LINE_LENGTH];
   static int in = 0;
   switch (c) {
      case '\n':                          /*   New line   */
                  line[in] = '\0';
                  printf("%s\n", line);
                  in = 0;
                  break;
      case '\f':                          /*   Formfeed   */
                  break;
      default:
                  if (in == LINE_LENGTH-1) {
                     line[in] = '\0';
                     printf("%s\n", line);
                     in = 0;
                  }
                  line[in++] = (char) c;
                  break;
   }
}
```

PostScript project—line—processing function

```
void proc_line(char *line)   /*   Process line   */
{
   if (ypos < BOTTOM_MARGIN) {
      ypos = TOP_MARGIN;
      if (xpos == LEFT_MARGIN)
         xpos = RIGHT_MARGIN;
      else {
         xpos = LEFT_MARGIN;
         /*   *******  New page required here   *******   */
      }
   }
   printf("%d %d moveto\n(", xpos, ypos);

   while (*line != '\0') {
      if ((*line == '(') || (*line == ')') || (*line == '\\'))
         putchar('\\');
      putchar(*(line++));
   }
   printf(") show\n");
   ypos -= LINE_HEIGHT;
}
```

PostScript project—complete solution

```
/*   Text to PostScript converter    */
/*   P. Jarvis            23/03/1991    */

#include <stdio.h>

#define LINE_LENGTH 80
#define TOP_MARGIN 536
#define BOTTOM_MARGIN 24
#define LEFT_MARGIN 30
#define RIGHT_MARGIN 424
#define LINE_HEIGHT 9

int main(int, char **);
void exit(int);
void proc_char(int);
void proc_line(char *);
void newpage(char *, int);

int xpos = RIGHT_MARGIN;
int ypos = BOTTOM_MARGIN - 1;
char *title;

int main(int argc, char *argv[])
{
   FILE *fh;
   int c;

   switch (argc) {
     case 1:
             fh = stdin;
             title = "Standard input";
             break;
     case 2:
             fh = fopen(argv[1], "r");
             if (fh == NULL) {
                fprintf(stderr, "Unable to open %s\n",
                        argv[1]);
                exit(1);
             }
             title = argv[1];
             break;
     default:
             fprintf(stderr, "Usage:   %s [filename]\n",
                     argv[0]);
             exit(2);
   }

   do {
     c = fgetc(fh);          /*   Get next character   */
     proc_char(c);           /*   Process it           */
   } while (c != EOF);        /*   Loop till end        */

   fclose(fh);
   return 0;
}
```

```c
void proc_char(int c)          /*   Process character   */
{
   static char line[LINE_LENGTH];
   static int in = 0;
   switch (c) {
     case EOF:
                 if (in != 0) {
                    line[in] = '\0';
                    proc_line(line);
                 }
                 if ((xpos != LEFT_MARGIN) ||
                     (ypos != TOP_MARGIN))
                    printf("showpage\n");
                 break;
     case '\n':
                 line[in] = '\0';
                 proc_line(line);
                 in = 0;
                 break;
     case '\f':
                 if (in != 0) {
                    line[in] = '\0';
                    proc_line(line);
                    in = 0;
                 }
                 ypos = BOTTOM_MARGIN - 1;
                 break;
     default:
                 if (in == LINE_LENGTH-1) {
                    line[in] = '\0';
                    proc_line(line);
                    in = 0;
                 }
                 line[in++] = (char) c;
                 break;
   }
}


void proc_line(char *line)  /*   Process line   */
{
   static int page = 1;

   if (ypos < BOTTOM_MARGIN) {
     ypos = TOP_MARGIN;
     if (xpos == LEFT_MARGIN)
       xpos = RIGHT_MARGIN;
     else {
       xpos = LEFT_MARGIN;
       newpage(title, page++);
     }
   }
   printf("%d %d moveto\n(", xpos, ypos);

   while (*line != '\0') {
     if ((*line == '(') || (*line == ')') || (*line == '\\'))
       putchar('\\');
     putchar(*(line++));
   }
   printf(") show\n");
   ypos -= LINE_HEIGHT;
}
```

```
void newpage(char *title, int page)
{
   if (page == 1)
     printf("%%!PS-Adobe\n");
   else
     printf("showpage\n");
   printf("90 rotate\n");
   printf("0 -595 translate\n");
   printf("20 20 moveto\n");
   printf("816 20 lineto\n");
   printf("816 550 lineto\n");
   printf("20 550 lineto\n");
   printf("closepath stroke\n");
   printf("418 20 moveto\n");
   printf("418 550 lineto\n");
   printf("stroke\n");
   printf("/Times-Roman findfont\n");
   printf("20 scalefont setfont\n");
   printf("20 555 moveto (%s) show\n", title);
   printf("730 555 moveto (Page %d) show\n", page);
   printf("/Courier findfont\n");
   printf("7.9 scalefont setfont\n");
}
```

34

Solving a maze

Uses:

The solution to this project will probably use the following features:

- passing arguments to the main function;
- two–dimensional arrays;
- macros;
- memory allocation;
- linked lists;
- using the ANSI.SYS device driver.

Introduction

While this may appear to have only a fun value, the algorithms required to solve a maze are very similar to those required by programs that design printed circuit board layouts. The idea of finding a route from one point to another depending on certain constraints is common to both. For this project only a simple maze will be used, small enough to be entirely displayed on the screen. On faster machines the program will need to be slowed down so that the solution process can be watched.

Getting started

The maze data can either be obtained as explained earlier, or can be entered manually as shown in Fig. 34.1. If entered manually please take extreme care as there is nothing more frustrating than trying to solve a maze that has no solution. Even a computer cannot do that! Ensure that the correct number of characters per line are entered: trailing spaces will be ignored. Do not use tab characters within the data.

Fig. 34.1   The maze to be solved

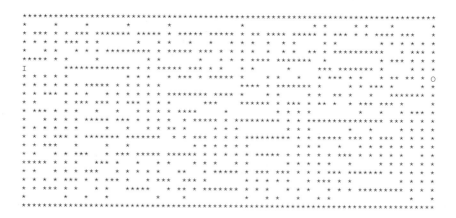

As usual, the first part to write is the main function which should open and close the maze data file, and exit. The file name can either be coded as a character string or, preferably, entered as a parameter to the program. Obviously all the checks should be made to ensure the file was opened correctly. Suitable error messages are a necessity.

Reading the maze

Next write a function to read the maze from the data file, but before the data can be read a decision is required as to how the maze should be stored in memory. Each point in the maze requires a value that indicates what part of the maze it is, for example a wall or a path. Also, when solving the maze, a linked list could be used to good effect; how will be demonstrated later. Thus it is suggested that the maze is stored in a structure of the following form:

```
struct cell {
  int cell;
    struct cell *next;
};
```

The cell value will be used to represent the contents of the maze element. Initially this will only be wall or path, but when solving the maze it will indicate from which direction the route entered the cell. Table 34.1 suggests values.

Table 34.1 Suggested maze cell representation

Value	Meaning
−1	Cell is part of a wall
0	Cell is unused
1	Route enters cell from above
2	Route enters cell from right
3	Route enters cell from below
4	Route enters cell from left
98	Start of route (IN)
99	End of route (OUT)

Initially represent the maze as a two–dimensional array. This array will be used by so many functions that it is probably worth making it global. Purists may disagree, but for the moment make it global. A suitable definition would be:

```
struct cell maze[80][23];
```

The function to read the maze data file should use `fgets` to read in each data line. While the data could be read one character at a time this would require testing for additional spaces and/or other random characters, at the end of each data line. Obviously the line buffer should be significantly longer than the expected 80 characters. Once read, the line length should be checked to ensure that at least the expected 80 characters are available. If not, there is some form of data error. Once happy the line has been read correctly, then the first 80 characters need processing. Each character needs mapping to the equivalent numeric value, that is, a space goes to zero, an asterisk to minus one, and so on, and this value requires saving in the corresponding element of the data array. The pointers in each cell should be set to `NULL`. The function should also save the row and column position of the input square as this will be needed when trying to find the route. The input and output cells are represented in the data file by the letters `I` and `O` respectively.

Writing the maze

Logically the next step must be to write the maze out. This will confirm that the maze has been read correctly and will also be required once a solution is found. In its simplest form this function will be the inverse of reading the data. Each value in the maze data will need mapping from its integer value to a character and this character written to the screen. Single character output will probably be simplest, rather than building an entire line and then printing it. It is likely that after writing the last character on a line, the DOS system will automatically move to the start of the next line. Inserting an extra `\n` will cause double spacing, which is not desired.

Using the ANSI.SYS device driver

Having done the above to prove that the data is being read correctly, expand the output function as follows. When solving the maze it would be nice to be able to watch the progress being made. This will require the ability to write a given character at a specific point on the screen. This can be done using the direct memory mapped technique as described in Chapter 18. Alternatively the ANSI display driver can be used. The second method will be described as it introduces yet another useful facility.

The DOS system allows device drivers to be installed. These are programs that receive data from a user program and forward it to a device. One such driver is ANSI.SYS. This is supplied as standard with DOS, but its use is optional. To check whether it is installed, list the file `config.sys` in the root directory of the disk from which the computer boots. If this file contains the following line, or something very similar, then the device driver is installed.

```
device=ansi.sys
```

If not, consult your DOS reference manual and install the ANSI device driver.

The `ansi.sys` device driver allows sequences of characters that conform to the X.3-64 ANSI standard to be correctly handled on the screen. These sequences allow for such things as moving to a particular position, clearing the screen and changing the colours. For example, the character sequence ESC [2 J will erase the screen. The sequence should contain no spaces and the character ESC is the escape code, which has the decimal value 27 (octal 33). Add the following line at the start of the maze output function and confirm that the screen gets cleared.

```
printf("\033[2J");
```

To write a character at a given line and column position requires the following escape sequence:

```
ESC [ y ; x H c
```

where:

```
ESC = Escape character (27 decimal)
  y = Line number (1 = top line)
  x = Column number (1 = left column)
  c = Required character
```

Appendix E summarizes some of the more useful escape sequences.

Write a function that takes a line and column position, and a character, and writes the character at the given position. Note how the escape sequence requires positions starting at one while C generally starts from zero. The function should take care of this difference.

Now modify the maze output function to call the single character output function just written, and check that all works as before.

Then one final change: modify the output to write the character with decimal value 219 (333 octal), instead of an asterisk, at each wall position. On an IBM/PC this character is a solid block graphic, which can be confirmed by referring to Appendix A. The result should look much more like a maze.

Visual improvements

If happy with the way the maze looks this part can be skipped as all that is done here is to tidy up the representation. Mazes are usually drawn using lines rather than blocks and the display is perfectly capable of handling this. There are a number of line drawing characters available that would produce a very nice looking maze, all that is needed is the following:

When drawing the maze, only cells that contain a wall value (−1) need processing. Each cell horizontally and vertically adjacent to the current cell has to be examined. On the basis of this, a suitably shaped character is selected. For example, if a wall cell had another wall cell above and to the right, then the required line drawing character for that cell would be 'L' shaped. Table 34.2 lists the line drawing characters and their decimal equivalents.

Table 34.2   IBM Line drawing characters

Offset	Value	Character
0	32	
1	179	│
2	196	─
3	192	└
4	179	│
5	179	│
6	218	┌
7	195	├
8	196	─
9	217	┘
10	196	─
11	193	┴
12	191	┐
13	180	┤
14	194	┬
15	197	┼

Why does Table 34.2 contain both blank and duplicate entries? (it's a clue). Supposing that one were examining the cells around the current cell which contains a wall value. Starting with a zero value, add one if the cell above is a wall, two for the cell to the right, four for the cell below and eight for the cell to the left. Then use the final value as an offset into Table 34.2 to determine the correct character to draw.

This technique of calculating some value and then looking up a result in a table, is known as 'table lookup'. Often by careful design of the data table programs can be speeded up. For example, it is common to build a table of sines or cosines when doing repetitive trigonomtrical calculations, to avoid repeating these calculations.

Modify the maze–writing routine to use this feature. If the table that maps offset to decimal character value is defined within a function, declaring it as static will avoid the array being initialized every time the function is called.. If all goes well your output should look something like that shown in Fig. 34.2.

Fig. 34.2 Improved maze display

Changing the maze representation

Up until this point a two–dimensional array has been used to represent the maze. While this is perfectly reasonable, a change is suggested for two reasons. Firstly, as explained earlier, it is better to allocate large amounts of memory at run time. This is especially true if the space required is not known until the program is run. Secondly this is a good place to demonstrate the use of macros. While neither point is a conclusive argument for the change the final solution does look neater. Often, but not always, a neat solution is best.

To make the change replace the maze array definition by a pointer. Use a different name, such as `mazep`. The definition should again be global, the following is suggested:

```
struct cell *mazep;
```

In the main program, use the function `malloc` to allocate enough memory for the maze. The following lines being a suggestion:

```
i = WIDTH * HEIGHT * sizeof(struct cell);
mazep = malloc(i);
```

Do not forget to check the return value from `malloc` and process any error condition correctly. Use informative error messages.

Now define the macro `maze` as shown below. This definition should be global, and must be before any function that uses it.

```
#define maze(x,y) mazep[(y) * WIDTH + x]
```

It is now possible to refer to a specific cell in the maze in three different ways. These are:

```
*(mazep + y * WIDTH + x)         as a pointer
mazep[y * WIDTH + x]             as an array
maze(x,y)                        as a macro
```

Change all occurrences of `maze[x][y]` to `maze(x,y)` throughout the entire program and then check that all works as before. We are now in a position to solve the maze.

Solving the maze

This is the fun part of the project and it is not as hard as may be thought. The technique recommended is that of trial and error. While rather rudimentary it is remarkably effective.

As a route is traced through the maze a record is needed of the cells passed. To do this each cell has an integral value saved in it, indicating from which direction the cell was entered. Once a complete route has been found these values can be used to retrace the route. Suggested cell values are listed in Table 34.1.

As an example, consider Fig. 34.3 which shows the various stages of solving a section of a maze. The starting point is shown in Fig. 34.3a where the entry is through the left edge, the required exit through the right edge, and wall cells are indicated using asterisks. Figure 34.3b shows the first stage where one cell is occupied. The value of four indicates the cell was entered from the left. Figure 34.3c shows the only possible next step. After this, two routes are possible, either up or down. Both routes must be explored. Follow through the remaining stages, checking that you follow the numbers entered.

Once the exit has been found, trace the route back, replacing each number by some other character, such as a dot. Having done this replace any cells still containing numbers with blanks and the result should be as shown in Fig. 34.3i.

Fig. 34.3 Stages in solving a section of maze

a

*	*	*	*	*
*				*
		*		
*		*		*

b

*	*	*	*	*
*				*
4		*		
*		*		*

c

*	*	*	*	*
*				*
4	4	*		
*		*		*

d

*	*	*	*	*
*	3			*
4	4	*		
*	1	*		*

e

*	*	*	*	*
*	3	4		*
4	4	*		
*	1	*		*

f

*	*	*	*	*
*	3	4	4	*
4	4	*		
*	1	*		*

g

*	*	*	*	*
*	3	4	4	*
4	4	*	1	
*	1	*		*

h

*	*	*	*	*
*	3	4	4	*
4	4	*	1	4
*	1	*	1	*

i

*	*	*	*	*
*	.	.	.	*
.	.	*	.	.
*		*		*

To implement this procedure only two functions are required: one to do the processing involved with occupying a cell, and the other to decide which cells to occupy. These functions will now be described in detail.

Occupying a cell

Occupying or marking a cell that forms part of a route involves several steps. Clearly the cell value must be set to indicate the direction of entry, but also two other actions are required. The display requires updating so that the solution can be watched, and also the cell must be marked for further processing as the adjacent cells may need to be checked.

The function will need two parameters. One is the address of the cell being occupied while the other is the direction from which the cell was entered. The function returns a true value if the exit cell is reached, otherwise a false value is returned. The following would be a suitable prototype for the function:

```
int occupy(struct cell *, int);
```

The first process in the function must be to check if the exit square has been reached. This can be recognized by the cell value being 99. If this is so, return from the function with a true value.

Otherwise further processing is required only if the required cell contains a zero value. Anything else would indicate that the cell was either a wall cell, or part of an existing route, that is, another route to the square has already been found. If the cell does contain a zero value, continue by first saving the given entry direction code as the new cell value. Then, to enable progress to be watched on the screen, the fact that the cell is now occupied needs recording. This can be done using the write character function produced earlier. The integer direction can be converted to a character by adding ' 0 '.

If p is the address of the current cell, then the row and column numbers (starting from zero) can be derived using:

```
col = (p - mazep)  %  WIDTH;
row = (p - mazep)  /  WIDTH
```

Write this much of the function and then test it as follows.

The maze input cell position is known: it should have been saved when the maze was read in. Determine the address of the cell to the right and call the occupy function using this address and the input direction of four. When the program is run the number four should be printed to the right of the input cell.

When this much works the final part of the occupy function requires adding. This part has to indicate that this cell needs to be processed by the second function whose job will be to examine adjacent cells.

While recursion might seem the correct approach, a linked list is easier to implement and is less demanding on memory. To implement a list, a pointer variable will be required to hold the address of the first cell in the list. This variable will be required by both this function and the second so make it global. The variable should be initialized to NULL as in:

```
struct cell *first = NULL;
```

Adding a cell to the start of the list is easiest and, if p is the address of the current cell, can be accomplished using:

```
p->next = first;
first = p;
```

Check you follow this process by drawing the linked list as it develops in the first few stages of Fig. 34.3.

Processing cells

The second function required to determine the maze solution takes cells from the linked list and calls the occupy function for each horizontally and vertically adjacent cell. Four calls to the occupy function will be required for each cell processed. To take the next cell from the list the following could be used:

```
struct cell *p;
p = first;
first = first->next;
```

Having got the address of the current cell use the occupy function to process the four adjacent cells. For example the cell above can be processed using:

```
occupy(p-WIDTH, 3);
```

Having written this much, test it by calling it from the main function after the first cell has been occupied. When the program is run two more numbers should appear in the maze, a one and a four.

If all is well the function must now be changed so that cells are taken from the list until either the exit is found, or the cell list becomes empty. The latter indicates that the maze has no solution. Enclose the calls to the occupy function within a loop which loops while the value of first is not null. Then if any of the four calls to occupy returns a true value, indicating the exit has been found, use a break statement to exit the loop.

Try this much and hopefully the entire maze will fill with numbers. If keen, the path can be manually retraced from the exit to ensure that a correct route has been found. Otherwise this can wait for the next section.

Retracing the maze

One final function and the program is complete. This function takes a cell address and retraces the route, replacing each cell value with the number 99. Once this has been done the maze is redrawn so that any value of 98 or 99 is drawn as an asterisk while any number in the range 0 to 4 is drawn as a space. The cell walls need not be redrawn as they should not have changed.

The maze is now solved! One final modification for those using faster computers. To slow down the solution so that progress can be watched, modify the character output function so that the character is written to the screen say twenty times. The increased output should slow the program enough to be watchable.

Maze solution—getting started

```
/*   Maze solving program - main  function   */
/*   P. Jarvis                     4/10/89   */

#include <stdio.h>

#define HEIGHT 23
#define WIDTH 80
#define EMPTY 0
#define WALL -1
#define IN 98
#define OUT 99
#define ABOVE 1
#define RIGHT 2
#define BELOW 3
#define LEFT 4

int main(int, char**);
void exit(int);

int main(int argc, char *argv[])
{
   FILE *fh;

   if (argc != 2) {
     fprintf(stderr, "Usage:   %s filename\n", argv[0]);
     exit(1);
   }

   fh = fopen(argv[1], "r");
   if (fh == NULL) {
     fprintf(stderr, "Unable to open %s\n", argv[1]);
     exit(2);
   }

   return 0;
}
```

Maze project—loading the maze

```
void load_maze(FILE *fh, int *xin, int *yin)
{
  char line[WIDTH+10];
  int i, j, k;

  for (i=0; i<HEIGHT; i++) {
    if (fgets(line, WIDTH+10, fh) == NULL) {
      fprintf(stderr, "Premature end of data\n");
      exit(3);
    }
    for (j=0; j<WIDTH; j++) {
      switch (line[j]) {
        case ' ':                  /*  Path          */
                k = EMPTY;
                break;
        case '*':                  /*  Wall cell     */
                k = WALL;
                break;
        case 'I':                  /*  Input cell    */
                k = IN;
                *xin = j;
                *yin = i;
                break;
        case 'O':                  /*  Output cell   */
                k = OUT;
                break;
        default:
                fprintf(stderr, "Illegal character - %c\n",
                         line[j]);
                exit(4);
      }
      maze[j][i].cell = k;
      maze[j][i].next = NULL;
    }
  }
}
```

Maze project—initial output function

```
void write_maze(void)
   int  i,  j,  k;

   printf("\033[2J");
   for (i=0;  i<HEIGHT;  i++) {
      for (j=0;  j<WIDTH;  j++) {
         switch(maze(j,i).cell) {
            case EMPTY:
                      k = ' ';
                      break;
            case WALL:
                      k = '\333';
                      break;
            case IN:
                      k = 'I';
                      break;
            case OUT:
                      k = 'O';
                      break;
            default:
                      k = '?';
         }
         printf("%c", k);
      }
   }
}
```

Maze solution—improved output function

```
void write_maze(void)
{
   static int shape[16] = { 32, 179, 196, 192,
                           179, 179, 218, 195,
                           196, 217, 196, 193,
                           191, 180, 194, 197};
   int i, j, k;

   printf("\033[2J");
   for (i=0; i<HEIGHT; i++) {
      for (j=0; j<WIDTH; j++) {
         if (maze[j][i].cell == WALL) {              /*  Wall cell  */
            k = 0;
            if ((i > 0) && (maze[j][i-1].cell == WALL))
               k += 1;
            if ((j < WIDTH-1) && (maze[j+1][i].cell == WALL))
               k += 2;
            if ((i < HEIGHT-1) && (maze[j][i+1].cell == WALL))
               k += 4;
            if ((j > 0) && (maze[j-1][i].cell == WALL))
               k += 8;
            write_char(j, i, shape[k]);
         }
      }
   }
   printf("\n");
}


void write_char(int x, int y, int c)
{
   printf("\033[%d;%dH%c", y+1, x+1, c);
}
```

Maze project—cell occupy function version one, add cell at start of list

```
int occupy(struct cell *p, int dir)    /*  Occupy  given  cell    */
{
   int row, col;

   switch (p->cell) {
     case OUT:                                      /*  Exit  found   */
              return TRUE;
     case EMPTY:                                    /*  Unused  cell  *,
              p->cell = dir;
              col = (p - mazep) % WIDTH;
              row = (p - mazep) / WIDTH;
              write_char(col, row, dir + '0');   /*  Output   */
              p->next = first;
              first = p;                       /*  Add  to  cell  list   */
   }
   return FALSE;
}
```

Maze project—cell occupy function version two, add cell at end of list

```
int occupy(struct cell *p, int dir)    /*  Occupy  given  cell    */
{
   int row, col;
   struct cell *ptr = first;

   switch (p->cell) {
     case OUT:                                      /*  Exit  found   */
              return TRUE;
     case EMPTY:                                    /*  Unused  cell  */
              p->cell = dir;
              col = (p - mazep) % WIDTH;
              row = (p - mazep) / WIDTH;
              write_char(col, row, dir + '0');   /*  Output   */
              if (first == NULL)
                 first = p;                  /*  First  cell   */
              else {
                 while (ptr->next != NULL) /* Find end   */
                    ptr = ptr->next;
                 ptr->next = p;             /*  Add  at  the  end   */
              }
              p->next = NULL;
   }
   return FALSE;
}
```

Maze project—process cell list

```
void proc(void)                    /*  Process cell list   */
{
   struct cell *p;

   while ((p=first) != NULL) {     /*  Get cell from list  */
      first = first->next;

      if (occupy(p-WIDTH, BELOW))
         break;
      if (occupy(p+1, LEFT))
         break;
      if (occupy(p+WIDTH, ABOVE))
         break;
      if (occupy(p-1, RIGHT))
         break;
   }

   if (p != NULL)
      retrace(p);
}
```

Maze project—retrace and display route found

```
void retrace(struct cell *p)          /*   Retrace maze route   */
{
   int i, j;
   static int dir[4] = {-WIDTH, 1, WIDTH, -1};
   while (p->cell != IN) {
      i = dir[p->cell - 1];
      p->cell = OUT;
      p += i;
   }

   for (i=0; i<WIDTH; i++) {
      for (j=0; j<HEIGHT; j++) {
         switch(maze(i,j).cell) {
            case ABOVE:
            case BELOW:
            case RIGHT:
            case LEFT:
                       write_char(i, j, ' ');
                       break;
            case IN:
            case OUT:
                       write_char(i, j, '*');
                       break;
         }
      }
   }
}
```

Maze project—complete solution

```
/*   Maze solving program - main function  */
/*   P. Jarvis                    4/10/89   */

#include <stdio.h>
#include <stdlib.h>

#define TRUE 1
#define FALSE 0
#define HEIGHT 23
#define WIDTH 80
#define EMPTY 0
#define WALL -1
#define IN 98
#define OUT 99
#define ABOVE 1
#define RIGHT 2
#define BELOW 3
#define LEFT 4
#define maze(x,y) mazep[(y) * WIDTH + x]

struct cell {
   int cell;
   struct cell *next;
};
struct cell *mazep;
struct cell *first = NULL;

int main(int, char**);                    /*   main function        */
void exit(int);                           /*   exit function        */
void load_maze(FILE *, int *, int *);     /*   Load maze            */
void write_maze(void);                    /*   Write maze           */
void write_char(int, int, int);           /*   Write character      */
int occupy(struct cell *, int);           /*   Occupy cell          */
void proc(void);                          /*   Process cell list    */
void retrace (struct cell *);             /*   Retrace found route   */

int main(int argc, char *argv[])
{
   FILE *fh;
   int xin, yin;                    /*   Maze input square     */

   if (argc != 2) {
      fprintf(stderr, "Usage:  %s filename\n", argv[0]);
      exit(1);
   }
   mazep = malloc(WIDTH * HEIGHT * sizeof(struct cell));
   if (mazep == NULL) {
      fprintf(stderr, "Unable to allocate space for maze\n");
      exit(3);
   }
   fh = fopen(argv[1], "r");
   if (fh == NULL) {
      fprintf(stderr, "Unable to open %s\n", argv[1]);
      exit(2);
   }

   load_maze(fh, &xin, &yin);
   write_maze();
   occupy(&maze(xin+1, yin), 4);
   proc();
   write_char(0, 23, ' ');
   getchar();
   return 0;
}
```

```
void load_maze(FILE *fh, int *xin, int *yin)
{
  char line[WIDTH+10];
  int i, j, k;

  for (i=0; i<HEIGHT; i++) {
    if (fgets(line, WIDTH+10, fh) == NULL) {
      fprintf(stderr, "Premature end of data\n");
      exit(3);
    }
    for (j=0; j<WIDTH; j++) {
      switch (line[j]) {
        case ' ':                    /*  Path   */
                  k = EMPTY;
                  break;
        case '*':                    /*  Wall cell   */
                  k = WALL;
                  break;
        case 'I':                    /*  Input cell   */
                  k = IN;
                  *xin = j;
                  *yin = i;
                  break;
        case 'O':                    /*  Output cell   */
                  k = OUT;
                  break;
        default:
                  fprintf(stderr, "Illegal character- %c\n",
                          line[j]);
                  exit(4);
      }
      maze(j,i).cell = k;
      maze(j,i).next = NULL;
    }
  }
}


void write_maze(void)
{
  static int shape[16] = {  32, 179, 196, 192,
                           179, 179, 218, 195,
                           196, 217, 196, 193,
                           191, 180, 194, 197};
  int i, j, k;

  printf("\033[2J");
  for (i=0; i<HEIGHT; i++) {
    for (j=0; j<WIDTH; j++) {
      if (maze(j,i).cell == WALL) {                /*  Wall cell  */
        k = 0;
        if ((i > 0) && (maze(j, i-1).cell == WALL))
          k += 1;
        if ((j < WIDTH-1) && (maze(j+1, i).cell == WALL))
          k += 2;
        if ((i < HEIGHT-1) && (maze(j, i+1).cell == WALL))
          k += 4;
        if ((j > 0) && (maze(j-1, i).cell == WALL))
          k += 8;
        write_char(j, i, shape[k]);
      }
    }
  }
  printf("\n");
}
```

```c
void write_char(int x, int y, int c)
{
   printf("\033[%d;%dH%c", y+1, x+1, c);
}



int occupy(struct cell *p, int dir)          /*   Occupy given cell   */
{
   int row, col;
   struct cell *ptr = first;

   switch (p->cell) {
      case 99:                                /*   Exit found   */
               return TRUE;
      case 0:                                 /*   Unused cell   */
               p->cell = dir;
               col = (p - mazep) % WIDTH;
               row = (p - mazep) / WIDTH;
               write_char(col, row, dir + '0');    /*   Output   */
               if (first == NULL)
                  first = p;                  /*   First cell   */
               else {
                  while (ptr->next != NULL) /*   Find end      */
                     ptr = ptr->next;
                  ptr->next = p;              /*   Add at the end   */
               }
               p->next = NULL;
   }
   return FALSE;
}




void proc(void)                     /*   Process cell list   */
{
   struct cell *p;

   while ((p=first) != NULL) {       /*   Get cell from list   */
      first = first->next;

      if (occupy(p-WIDTH, BELOW))
         break;
      if (occupy(p+1, LEFT))
         break;
      if (occupy(p+WIDTH, ABOVE))
         break;
      if (occupy(p-1, RIGHT))
         break;
   }

   if (p != NULL)
      retrace(p);
}
```

```
void retrace(struct cell *p)
{
   int i, j;
   static int dir[4] = {-WIDTH, 1, WIDTH, -1};
   while (p->cell != IN) {
      i = dir[p->cell - 1];
      p->cell = OUT;
      p += i;
   }

   for (i=0; i<WIDTH; i++) {
      for (j=0; j<HEIGHT; j++) {
         switch(maze(i,j).cell) {
            case ABOVE:
            case BELOW:
            case RIGHT:
            case LEFT:
                    write_char(i, j, ' ');
                    break;
            case IN:
            case OUT:
                    write_char(i, j, '*');
                    break;
         }
      }
   }
}
```

35

Processing TIFF files

Uses

This project requires the use of the following techniques:

- bit manipulation;
- binary file input;
- random file positioning.

Introduction

TIFF (tagged image format file) is one of many pseudo–standard ways a graphical image can be stored. Most image scanners can produce this format of file, and there are many programs that can display, or incorporate them.

The data for this section are shown in Fig. 35.1, which was created by scanning an electrocardiograph trace (i.e. the output from a heart monitor) using a flat bed scanner. The idea is now to read this file and determine the x and y co-ordinates of each point along the line. This could, for example, be used to calculate the area under the trace, which can then be related to the blood flow. Although this project does not produce a stunning result, it does demonstrate a number of useful points for handling data.

Fig. 35.1   Electrocardiograph (ECG)

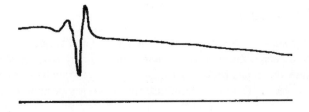

TIFF definition

A full definition of TIFF is beyond the scope of this text; only sufficient details will be covered to enable the project to be completed. One major item not covered is that TIFF files can include data compression techniques. The supplied data is not compressed and so this complication is avoided.

A TIFF file consists of a header, one or more directories, and one or more data blocks. The header is of a fixed size while the directories contain a variable number of entries, each of which is a fixed size. The data size is obviously variable. The three parts are defined as follows.

File header format

The header is formed of the first eight bytes in the file. The first two bytes indicate the order in which data is stored. If two data bytes are read (a and b) and these represent a single number, then the numeric value can be either of the following:

```
a × 256 + b
b × 256 + a
```

If the first two bytes of the header both contain the letter M all subsequent numbers use the first interpretation (i.e. high byte first). If they both contain the letter I then the order is low byte then high byte. Any other values are illegal.

The next two bytes of the header form a version number. Note that the order of the bytes will have been determined from the initial two bytes and this ordering must be taken into account on this, and all subsequent, numeric input. If the version number is not 42 then it is illegal.

The next four bytes end the header, and these bytes (whose order is again specified by the initial two bytes of the file) give the offset from the start of the file of the first directory block.

Directory structure

Each directory starts with a two–byte value specifying the number of entries in the directory. Each of these entries consists of two two–byte numbers, known as the tag and the type, followed by two four–byte numbers, known as the count and the value. The tag determines how the entry is interpreted and thus which other fields are meaningful. Appendix F lists the entries found in the supplied data but remember that this is by no means a complete list, neither are all the terms used explained. Those values required for the project will be described as necessary.

The type field specifies how the value field is to be processed. There are five possible formats and these are shown in Table 35.1.

Table 35.1   Type codes

Type	Meaning
1	8–bit unsigned integer
2	ASCII string (nul terminated)
3	16–bit unsigned integer
4	32–bit unsigned integer
5	Rational number

Where the required value can be entirely contained within four bytes, then the value field contains that value. If the required value is more than four bytes, then the value field contains a file offset that defines where the required data is within the file. Type 5 date requires two 4 byte integers and divides the first by the second, thus producing a real value. This type code always requires the offset to be given. For the other types, where the required value fits within the four–byte field, then the value is left justified; the additional trailing bytes being ignored.

The count field specifies the number of data bytes. For type 2 data (ASCII string) it specifies the number of characters including the terminating zero byte.

Image structure

The image structure is defined by the corresponding directory. For the supplied data it has the following form. Each byte of data contains eight image pixel values. Thus each pixel is either black or white. In the data a zero value indicates white, while a value of one indicates black. The image was scanned starting at the lower left corner with each scan line being in the vertical direction. Subsequent scans moving to the right. Each scan line is a whole number of bytes, and thus the last byte of each scan line may contain unused bits.

Getting started

The required program needs to process a data file. The name of this file needs to be specified either as a parameter or by entering from the keyboard. Decide on the preferred method (the solution supplied uses program parameters) and start by writing a main function that opens and, if successful then closes, the specified data file. Note that the file must be opened in binary mode. Check the program behaves correctly if no filename is specified, and if the file cannot be opened (e.g. it does not exist). It cannot be over–emphasized that meaningful error messages should be issued whenever an error condition occurs. When happy all is well progress to reading the header.

Reading the header

The most important two bytes of the header are the first two. These are either
II or MM. Read these two bytes, using `fgetc`, verify that they are correct, and
set a flag to true for II, false for MM.

Next a function is required that reads a numeric value from the data file.
Apart from the image data, all numeric values are one, two, or four bytes long.
The byte order is determined from the flag set from the first two bytes of the
header. Write a function that takes three parameters, one a `FILE *`, and two of
type `int`. The function should return a `long int` value. The three parameters
specify the file handle, how many bytes to read (one, two, or four), and the
byte order. The byte order parameter will be the flag variable set earlier.

Use this function to read the version number (two bytes) and the file offset
of the first directory (four bytes) and then print these values. If, for the
supplied data, the version number is not 42, or the directory offset not 8, then
there is an error. Once working do not bother to output the values anymore;
simply ignore the version and save the directory offset. Processing the
directory comes next.

However, before looking at the directory, take a look at the supplied
solution to this part (p. 225). Note particularly the value used in the `switch`
statement. A common technique when testing two, or more, values is to
combine them into one value and do a single test. Thus the first byte read is
multiplied by 256 and the second byte added. The resulting 16–bit value is
tested against the two defined symbols (II and MM). It would be very tempting
to combine the two calls to fgetc into one statement, such as:

```
switch (fgetc() * 256 + fgetc()) {
```

This is not correct. Although the multiplication will be done before the
addition, there is nothing in the C standard which defines the order in which
functions are called. Thus two possible interpretations are possible. The first
option is what would be expected, one byte is read, this value is multiplied by
256, and this is added to the next byte read. However, it is also perfectly valid
for the program to read one byte from the file, save that value somewhere, read
another byte, multiply this by 256 and add the saved byte, thus producing a
different result. To ensure correct evaluation order the two function calls have
been placed in separate statements.

Processing the directory

Next write a function to process a directory. This will require the file handle
(of type `FILE *`), the offset to the start of the directory (as a `long int`), and
the byte ordering flag (as an `int`). Use the `fseek` function to position the file
at the start of the required directory. It could well be worth writing a function
that uses `fseek` to position the file at a given offset. This function should exit

the program if the `fseek` fails, and should also return the original file offset (obtained using `ftell`) if all is well. Using such a function will reduce the amount of checking required, and will produce easier to read code.

After positioning the file at the start of the directory, initially just print out the directory entries. The directory starts with a two–byte value that specifies the number of directory entries which follow. After the last directory entry is a four–byte file offset to the next directory. If this offset is zero then there are no more directories. Each directory entry consists of a tag (2 bytes), a type (2 bytes), a count (4 bytes), and a value (4 bytes). Remember that how the value field is interpreted depends on the type code. Some value fields will contain file offsets which will require reading a different section of the file to get the true value and then restoring the original file position. Write a loop which reads and prints every directory in the file. Do not expect more than one directory to be printed, as the supplied data has only one. When the output matches that shown in Table 35.2 then proceed to the next stage.

Table 35.2   Expected directory entries

255	3	1	1
256	3	1	274
257	3	1	522
258	3	1	1
259	3	1	1
262	3	1	1
263	3	1	1
266	3	1	1
273	4	1	246
274	3	1	1
277	3	1	1
278	4	1	522
279	4	1	18270
280	3	1	0
281	3	1	1
282	5	1	150.00
283	5	1	150.00
284	3	1	1

Simplifying the directory processing

When it comes to processing the data then, for the supplied data, just four directory entry tag values need be processed. The required tag values are 256 (image width), 257 (image height), 273 (offset to data), and 279 (total data byte count). The first two tags are of type three (i.e. require a two–byte value field) while the last two are of type four (i.e. require a four–byte value field).

To simplify the program re-write the previous directory reading function so that all directory entries, other than the required four, are ignored. Thus the code that handled the rational number input and processing ASCII strings can be removed. Save the image height, width, data offset, and total size in suitably named variables and initially just print them out. The values obtained should match those given in Table 35.2.

When the displayed values are as expected then a data–processing function needs writing that takes these four values as parameters. The intention of this function is to write out the distance between the baseline and the ECG trace. The value can be written in dots (each dot is specified in the directory header as being 1/150 of an inch).

Processing the data

The first point to establish is the data orientation. This is specified by one of the directory entries, but for handling the supplied data all one needs to know is that the longest dimension is the height. The scanning direction, in this case the width, is the shorter dimension, and the baseline is reached first, i.e. is to the left of the real trace. In other words if one rotates Fig. 35.1 clockwise by ninety degrees so that the base line is on the left side, then the data is read starting at the top left corner from left to right, then top to bottom, just like a television picture.

The first actions the data processing function should take are to save the current file position, and move to the start of the data. The number of scan lines is known, as this is the image height, and the number of pixels on each line is also known, as this is the width. The number of bytes per scan line can be calculated by dividing the total number of bytes in the image by the image height.

Knowing the number of bytes in each scan line allocate, using `malloc`, enough space to hold one line. Then set up a loop that reads in every scan line using the `fread` function. Obviously the return values from both standard functions should be checked and a suitable message printed if required. Finally, reset the file position back to the value saved on entry. The next section will deal with processing each line but check this much works before progressing.

Processing each line

The previous function will read in 522 scan lines and will pass each line to a line handling function. This function will return the distance, in dots, between the base line and the ECG trace. In order to do this it will be necessary to search the scan line for the position of the base line, then continue the search to find the trace, and finally return the difference. As both the base line and the trace may be several dots wide, it would be a good idea to find both edges of each line and take the midpoint as the true value. To achieve this will require the ability to access any particular bit in the scan line, and this probably warrants a function.

Write such a function that should take two parameters. The first is a character array while the second is an integer. The first parameter will be the scan line read in and the integer will be the pixel position to test, starting with zero as the first point. The first character in the array will contain the first eight pixels, numbered zero on the left to seven on the right. The next character will contain bits eight to fifteen, and so on. The function should return the value of the bit either zero or one. Write this function, possibly with its own simple test program, and check that it works.

Once a bit–testing function as described above is available, return to the line–processing function and proceed as follows. Search from the first position until the first bit position is found that contains a one. Remember this position and then continue the search from this point until the next zero bit is found. The distance between this point and the previous will be the width of the base line. Calculate the average and print it out. Just for the moment leave the processing at this and test it.

When the program is run the base line position should be around the sixty mark but will vary slightly. This is because when the image was scanned the trace was not exactly square on the scanner. By using a base line this slight distortion can be determined and its effect overcome.

When this much is working extend the line–processing function by adding a second search section that locates the trace line and then returns the difference between the trace and the base line. Next it should be possible to output a set of co-ordinate pairs where the first number in each pair is the scan line and the second number is the separation distance. These values could be plotted or processed to obtain area information if required, but for now this project has covered the main points of random file access, data structure handling and bit manipulations. Any extensions are left to the reader.

It would be possible, for example, to plot the corrected trace on the screen but this requires information that is not covered for another two chapters. However, there is no reason why this project should not be continued later.

TIFF—getting started

```
/*  TIFF file processing - Getting started       */
/*  P. Jarvis                    07/02/1992       */

#include <stdio.h>

int main(int, char **);
void exit(int);

int main(int argc, char *argv[])
{
   FILE *fh;

   if (argc != 2) {
     fprintf(stderr, "Usage:  %s filename\n", argv[0]);
     exit(1);
   }

   fh = fopen(argv[1], "rb");
   if (fh == NULL) {
     fprintf(stderr, "Unable to open %s\n", argv[1]);
     exit(2);
   }

   fclose(fh);
   return 0;
}
```

TIFF—reading the header

```
/*   TIFF file processing - reading the header    */
/*   P. Jarvis                        07/02/1992    */


#include <stdio.h>

#define TRUE 1
#define FALSE 0
#define II 'I' * 256 + 'I'
#define MM 'M' * 256 + 'M'


int main(int, char **);
void exit(int);
long getnum(FILE *, int, int);


int main(int argc, char *argv[])
{
   FILE *fh;
   int byte_order;               /*   true if low byte first   */
   int i, j;

   if (argc != 2) {
      fprintf(stderr, "Usage:   %s filename\n", argv[0]);
      exit(1);
   }

   fh = fopen(argv[1], "rb");
   if (fh == NULL) {
      fprintf(stderr, "Unable to open %s\n", argv[1]);
      exit(2);
   }

/*   Read header   */

   i = fgetc(fh);
   j = fgetc(fh);
   switch (i * 256 + j) {
      case II:                          /*   Low byte first   */
              byte_order = TRUE;
              break;
      case MM:                          /*   High byte first   */
              byte_order = FALSE;
              break;
      default:
              fprintf(stderr, "Illegal header format\n");
              exit(3);
   }

   printf("Version number   %ld\n", getnum(fh, 2, byte_order));
   printf("Directory offset %ld\n", getnum(fh, 4, byte_order));


   fclose(fh);
   return 0;
}
```

```
long getnum(FILE *f, int num, int order)
{
    long l = 0;
    int i, shift, incr;
    if (order) {                           /*  If low byte first  */
        shift = 0;
        incr = 8;
    }
    else {                                 /*  If high byte first  */
        shift = (num - 1) * 8;
        incr = -8;
    }
    for (i=0; i<num; i++) {
        l += (long) fgetc(f) << shift;
        shift += incr;
    }
    return l;
}
```

TIFF—processing directories

```
long proc_directory(FILE *f, long offset, int order)
{
    int c, i, num;
    long tag, type, count, value, mark;
    double rational;

    myfseek(f, offset);
    num = (int) getnum(f, 2, order);
    for (i=0; i<num; i++) {
        tag = getnum(f, 2, order);
        type = getnum(f, 2, order);
        count = getnum(f, 4, order);
        printf("%5ld%5ld%5ld    ", tag, type, count);
        switch ((int) type) {
          case 1:                                 /*  Single byte  */
                value = getnum(f, 1, order);
                getnum(f, 3, order);
                printf("%ld\n", value);
                break;
          case 2:                                 /*  ASCII string    */
                if (count <= 0)
                   break;
                if (count <= 4) {
                   while ((c=fgetc(f)) != '\0')
                      printf("%c", fgetc(f));
                   getnum(f, 4 - (int) count, order);
                }
                else {
                   value = getnum(f, 4, order);
                   mark = myfseek(f, value);
                   while (count-- != 1L)
                      printf("%c", fgetc(f));
                   myfseek(f, mark);
                }
                printf("\n");
                break;
          case 3:                                 /*  Two byte value  */
                value = getnum(f, 2, order);
                getnum(f, 2, order);
                printf("%ld\n", value);
                break;
          case 4:                                 /*  Four byte value */
                value = getnum(f, 4, order);
                printf("%ld\n", value);
                break;
          case 5:                                 /*  Rational value  */
                value = getnum(f, 4, order);
                mark = myfseek(f, value);
                value = getnum(f, 4, order);
                rational = (double) value / (double) getnum(f, 4, order);
                myfseek(f, mark);
                printf("%.2f\n", rational);
                break;
        }
    }

    return getnum(f, 4, order);      /*  Offset to next directory  */
}
```

```
long myfseek(FILE *f, long offset)
{
    long l;
    l = ftell(f);
    if ((l == -1L) || (fseek(f, offset, SEEK_SET) != 0)) {
        perror("File positioning error");
        exit(3);
    }
    return l;
}
```

TIFF—accessing individual bits (test program)

```
#include <stdio.h>

int main(void);
int getbit(unsigned char *, int);

int main(void)
{
    static unsigned char data[] = {0x03, 0x44, 0x1F};
    int i;
    for (i=0; i < 24; i++)
        printf("%d", getbit(data, i));
    printf("\n");
    return 0;
}

int getbit(unsigned char *p, int bit)
{
    int i;
    i = p[bit / 8];
    return (i >> (7 - bit % 8)) & 1;
}
```

TIFF—processing data bytes

```
int proc_data(FILE *f, long height, long width, long nbytes,
              long offset)
{
  int i, nb;
  unsigned char *ptr;
  long mark;
  mark = myfseek(f, offset);
  nb = (int) (nbytes / height);
  ptr = malloc(nb);
  if (ptr == NULL) {
    fprintf(stderr, "Unable to allocate memory for data\n");
    exit(4);
  }
  for (i=0; i < (int)height; i++) {
    if (fread(ptr, 1, nb, f) != nb) {
      fprintf(stderr, "Error reading data block\n");
      exit(5);
    }
    printf("%5d %5d\n", i, proc_line(ptr, (int) width));
  }
  free(ptr);
  myfseek(f, mark);
  return 0;
}


int proc_line(unsigned char *line, int width)
{
  int i, j, base;
  for (i=0; i < width; i++) {                      /*  Base line start   *
    if (getbit(line, i) == 1) {
      for (j=i+1; j < width; j++) {                /*  Base line end   */
        if (getbit(line, j) == 0) {
          base = i + (j - i) / 2;
          for (i=j+1; i < width; i++) {            /*  Trace start   */
            if (getbit(line, i) == 1) {
              for (j=i+1; j< width; j++) {  /*  Trace end   */
                if (getbit(line, j) == 0)
                  return i + (j - i) / 2 - base;
              }
            }
          }
        }
      }
    }
  }
  return -1;
}


int getbit(unsigned char *p, int bit)
{
  int i;
  i = p[bit / 8];
  return (i >> (7 - bit % 8)) & 1;
}
```

TIFF—complete solution

```c
/*   TIFF file processing - Complete solution   */
/*   =========================================   */

/*   P. Jarvis                    07/02/1992   */

#include <stdio.h>
#include <stdlib.h>

#define TRUE 1
#define FALSE 0
#define II 'I' * 256 + 'I'
#define MM 'M' * 256 + 'M'

int main(int, char **);
void exit(int);
long getnum(FILE *, int, int);
long proc_directory(FILE *, long, int);
long myfseek(FILE *, long);
int proc_data(FILE *, long, long, long, long);
int proc_line(unsigned char *, int);
int getbit(unsigned char *, int);

int main(int argc, char *argv[])
{
  FILE *fh;
  int byte_order;            /*   true if low byte first   */
  int i, j;
  long dir_offset;

  if (argc != 2) {
    fprintf(stderr, "Usage:   %s filename\n", argv[0]);
    exit(1);
  }

  fh = fopen(argv[1], "rb");
  if (fh == NULL) {
    fprintf(stderr, "Unable to open %s\n", argv[1]);
    exit(2);
  }

/*   Read header   */

  i = fgetc(fh);
  j = fgetc(fh);
  switch (i * 256 + j) {
    case II:                 /*   Low byte first   */
            byte_order = TRUE;
            break;
    case MM:                 /*   High byte first   */
            byte_order = FALSE;
            break;
    default:
            fprintf(stderr, "Illegal header format\n");
            exit(3);
  }

  getnum(fh, 2, byte_order);
  dir_offset = getnum(fh, 4, byte_order);
  while (dir_offset != 0L)
    dir_offset = proc_directory(fh, dir_offset, byte_order);
  fclose(fh);
  return 0;
}
```

```
long getnum(FILE *f, int num, int order)
{
  long l = 0;
  int i, shift, incr;
  if (order) {                        /*  If low byte first  */
    shift = 0;
    incr = 8;
  }
  else {
    shift = (num - 1) * 8;
    incr = -8;
  }
  for (i=0; i<num; i++) {
    l += (fgetc(f) << shift);
    shift += incr;
  }
  return l;
}


long proc_directory(FILE *f, long offset, int order)
{
  int i, num, skip;
  long width, height, nbytes, data_offset;
  long tag, type, count;

  myfseek(f, offset);
  num = (int) getnum(f, 2, order);
  for (i=0; i<num; i++) {
    tag = getnum(f, 2, order);
    type = getnum(f, 2, order);
    count = getnum(f, 4, order);
    skip = 4;
    switch ((int) tag) {
      case 256:                       /*  Image width       */
        width = getnum(f, 2, order);
        skip = 2;
        break;
      case 257:                       /*  Image height      */
        height = getnum(f, 2, order);
        skip = 2;
        break;
      case 273:                       /*  Data offset       */
        data_offset = getnum(f, 4, order);
        skip = 0;
        break;
      case 279:                       /*  Byte count        */
        nbytes = getnum(f, 4, order);
        skip = 0;
        break;
    }
    if (skip != 0)
      getnum(f, skip, order);
  }
  proc_data(f, height, width, nbytes, data_offset);
  return getnum(f, 4, order);         /*  offset to next directory  */
}
```

```
long myfseek(FILE *f, long offset)
{
   long l;
   l = ftell(f);
   if ((l == -1L) || (fseek(f, offset, SEEK_SET) != 0)) {
     perror("File positioning error");
     exit(3);
   }
   return l;
}

int proc_data(FILE *f, long height, long width, long nbytes,
             long offset)
{
   int i, nb;
   unsigned char *ptr;
   long mark;
   mark = myfseek(f, offset);
   nb = (int) (nbytes / height);
   ptr = malloc(nb);
   if (ptr == NULL) {
     fprintf(stderr, "Unable to allocate memory for data\n");
     exit(4);
   }
   for (i=0; i < (int)height; i++) {
     if (fread(ptr, 1, nb, f) != nb) {
       fprintf(stderr, "Error reading data block\n");
       exit(5);
     }
     printf("%5d %5d\n", i, proc_line(ptr, (int) width));
   }
   free(ptr);
   myfseek(f, mark);
   return 0;
}

int proc_line(unsigned char *line, int width)
{
   int i, j, base;
   for (i=0; i < width; i++) {
     if (getbit(line, i) == 1) {                    /*  Base line start  */
       for (j=i+1; j < width; j++) {
         if (getbit(line, j) == 0) {                /*  Base line end  */
           base = i + (j - i) / 2;
           for (i=j+1; i < width; i++) {
             if (getbit(line, i) == 1) {            /*  Trace start  */
               for (j=i+1; j< width; j++) {
                 if (getbit(line, j) == 0)          /*  Trace end  */
                   return i + (j - i) / 2 - base;
               }
             }
           }
         }
       }
     }
   }
   return -1;
}

int getbit(unsigned char *p, int bit)
{
   int i;
   i = p[bit / 8];
   return (i >> (7 - bit % 8)) & 1;
}
```

36

Text windowing system

Uses:

This is a more adventurous project and uses the following features:

- structures;
- linked lists;
- memory allocation;
- pointers and pointers to pointers.

Introduction

The idea of this project is to produce a mechanism for a program to write text to a number of different rectangular areas of the screen, called windows. Note that it is not the intention to develop a complete windowing system with mouse interface and pull down menus. This is somewhat beyond our scope at the moment.

The sort of use for the proposed facility would be to enable a program to produce output in more than one window. Each window could be as large as the screen and there would be some way of selecting the currently viewed window. So, for example, the main output could be written to one window while some debugging information was written to another, and program input entered via a third.

The functionality provided by this project will enable windows to be created, displayed, text written to these windows, and for the order of the windows to be changed. Thus, when windows overlap, a given window can be moved so that it is on top (i.e. completely visible) or moved to the bottom (i.e. could be totally covered by other windows).

Getting started

The first requirement of any windowing system is a fast method of writing to the screen. If we confine the project to just text windows then the screen can be updated by direct memory access, as was described in Chapter 16 when discussing pointers.

Just to refresh this topic, the display screen is memory mapped. This means that there is an area of memory, containing the screen contents, which can be directly addressed from within a program. Changing the correct parts of this memory will automatically update the display. The display memory uses two bytes per displayed character: one holds the ASCII code for the character, while the other holds the required attributes. The attributes specify the required colour (if available), intensity, and whether the character is flashing or not. As each displayed character requires two separate values a structure is in order, the following being a suggestion:

```
struct screen {
    unsigned char c;    /*  Displayed character  */
    unsigned char a;    /*  Required attributes  */
};
```

The next requirement is to determine the display address. The address differs with screen mode but this can be determined by either using a BIOS request as was described in Chapter 29, the first exercise being just what is required here. The other rather frowned upon method of determining the display mode is to examine a specific byte in memory. Segment 0x40, offset 0x49 contains the current display mode. The code shown in Fig. 36.1 uses this byte to determine the display address. Note that it assumes that the screen mode is text and does not check this, neither does it check the screen size.

Enter the code shown in Fig. 36.1 as is, and check it works. Next create two global variables, one to hold the number of lines on the screen, and the other to hold the number of columns. Set these variables from within the function that determines the screen memory address. These values will depend on the current screen mode in use and are given in Appendix C. The program shown in Fig. 36.1 assumes the screen size to be 80 by 25 when it clears the screen. Modify the code to use the size determined from the screen mode.

One final check the function should do is to return some error condition if the screen is not in a text mode. As the function returns a pointer, the standard way of returning an error condition would be to return a NULL value.

Window structure

Now we know how to update the screen the next step must be to decide what information is required for each window that is displayed. The position and size of the window are obvious, but there are a few more items required.

For example, if part of the window is obscured and text is written to that window, and then the window is subsequently raised so that it becomes completely visible, the previously obscured text must be displayed. This requires all the text for a given window to be stored, whether or not it is being displayed. It would also be nice to have a name for the window that would form part of the display.

Fig. 36.1   Determining screen mode

```
#include <stdio.h>

struct screen {
  unsigned char c;
  unsigned char a;
};

int main(void);
struct screen _far *getadr(void);

int main(void)
{
  int i;
  struct screen _far *sptr;
  sptr = getadr();
  for (i=0; i<80*25; i++)
    sptr[i].c = 'A';
  getchar();
  return 0;
}

struct screen _far *getadr(void)
{
  unsigned char _far *p;
  p = (unsigned char _far *) 0x00400049;
  if (*p == 7)
    return((struct screen _far *) 0xB0000000);
  else
    return((struct screen _far *) 0xB8000000);
}
```

Two other useful variables are a unique window number and a text offset. Both of these will be described later. Figure 36.2 shows a suitable structure definition.

The height and width of a window describe the size of the text area to be displayed. It may be that a border is drawn round a window, in which case the height and width do not include the border. Similarly the window position, which is referenced by the top left corner, does not include any border if one is drawn.

The two character pointers address the window text and window name respectively. As the size of neither is known at compile time the addresses must be assigned at run time using the function malloc (see Chapter 25 for details).

The remaining two variables, both pointers, will again be described later. For now just initialize their values to NULL and ignore them.

Fig. 36.2 Window structure definition

```
struct window {
    int height;
    int width;
    int xpos;
    int ypos;
    int num;
    int offset;
    unsigned char *text;
    unsigned char *name;
    struct window *prev;
    struct window *next;
};
```

Creating windows

Write a function to create a window structure. This will require the window's size, position, and a name. A suitable call would then be:

```
wcreate(6, 5, 30, 10, "Window 1");
```

This would create a window of 10 lines each of 30 characters. The window would be positioned at column 6, line 5 relative to the top left of the screen. Note that lines and columns both start at zero. The window would have the name 'Window 1'.

When the windows are displayed it would be nice to have a one character wide border as an outline. The IBM line drawing characters would make things look even better. To allow for this the window–create function should check that any window requested fits within the screen when a surrounding border is added. If this is done here then all subsequent functions can just assume there is room and skip any testing. Therefore windows must not be defined as starting on the top line or in the left column. Remember the position of the window is that of the text; the border is outside. Also check that the window plus border does not fall off the edge of the screen. Thus, for an 80 by 25 display, the largest text window possible would be 78 columns by 23 lines. Return a NULL value if the requested window will not fit within the screen.

Having checked that the window will fit, the next action required by the create function will be to allocate space for the window structure, the following being a good starting point:

```
struct window *wptr;
wptr = malloc(sizeof(struct window));
```

The function will eventually return this address as its value. Should an error occur during the allocation, for example no memory is available, then a NULL value should be returned.

If the structure is allocated successfully then save the size and position values therein. Then allocate enough space to hold the text. This space should be preset to the space character. Similarly, space needs allocating for the window name. Use strlen to determine the required size and don't forget to add one for the terminating zero byte. Copy the given name into the allocated area, possibly using strcpy. Finally set both the text offset variable and the window number to zero, and the remaining pointers to NULL. These later variables will be assigned in due course.

Having written this function, test that it works as far as possible but don't expect it to draw anything on the screen, that bit comes next.

Displaying windows

At last here is the part where we actually start to see some output on the screen. A function is required to display a window given the address of the window structure. Two routines are probably a good idea, one to draw the window border and other unchanging parts and the other to write the text within the window. The second function is called each time the text within a window is modified while the first only requires calling when the window is created and whenever a previously covered part of the window becomes visible.

This section describes only the first function. It requires the window structure address as a parameter so the following would be a suitable function prototype:

```
void wdisplay(struct window *);
```

The window size and position are known so the location of the border can be calculated. The border should be constructed using the line drawing characters available in the IBM character set. Either refer to Table 34.2 or Appendix A.

First draw a row of horizontal line characters across the top and bottom of the window (outside the text area). To do this, declare two screen pointers (say p and q), one being the address in screen memory of the top left corner of the border area, and the other being the address of the bottom left corner. Then use these pointers as if they were arrays and set screen elements p[1].c to p[width].c and q[1].c to q[width].c to the horizontal line character.

Having set the horizontals, fill in the four corners (p[0].c, q[0].c, p[width+1].c and q[width+1].c). Finally draw the verticals. If feeling adventurous, insert the window name centrally in the top border.

When this much compiles, test it by creating two overlapping windows and calling the display function for each. Figure 36.3 shows a suitable test program. Hopefully the last window drawn will have a complete border, while the first will be partially covered.

Fig. 36.3   Testing the display function

```
int main(void)
{
    struct screen far *sptr;
    struct window *wptr;
    sptr = getadr();
    if (sptr == NULL)
        exit(1);
    wptr = wcreate(5, 5, 30, 10, "window 1");
    if (wptr != NULL)
        wdisplay(wptr);
    wptr = wcreate(7, 9, 30, 10, "window 2");
    if (wptr != NULL)
        wdisplay(wptr);
    getchar();
    return 0;
}
```

Displaying text

The function to display the text within a window is required next. This will use one of the previously unexplained variables in the window structure. Consider the problem of scrolling a new line into a window. The text is stored in a character array within the window structure. If the text were always stored so that the top line in the window was the first bytes in the text buffer then to implement a scroll up, all the characters in the text buffer would require moving up the buffer by one line length. This could be a rather time–consuming process.

A far better method would be to have an offset variable that indicated at what point in the buffer the top line of the display was stored. Then to implement a scroll all that is required is for this offset value to be incremented by one line length. Obviously when the offset value reaches the end of the text buffer it is wrapped around to the start.

The process of writing the text to the screen thus requires determining the position in the text buffer where the first character of the top line is stored. Starting at this point copy one line to the display. Continue for all remaining lines but if the end of the text buffer is reached then start from the beginning of the buffer.

If this is not yet clear maybe the description of the required function will help.

The function will require the address of the window structure passed to it, so a suitable function prototype would be:

```
void wshow(struct window *ptr);
```

A local variable, say `i`, of type `int` is required to specify which character is the next to be read from the text buffer in the window structure. This variable would initially be set to the value of the variable `offset` in the window structure.

Having determined the position of the text on the display copy the characters from the window text buffer, starting at position `i`, until one line has been written. At the end of the line check if the variable `i` has passed the end of the text buffer, if so reset it to zero, that is, the start of the buffer. Determine the screen address of the next line and copy this. Continue until all lines have been written.

Another minor modification that might help would be to temporarily change the create function so that instead of presetting the text buffer to spaces some printable character was used. If really feeling clever use a different character for each call to the create function, a static variable could be used here.

Modify the window display function so that its last statement is a call to the show text function. Thus the display function will draw the border and optional name and then call the show function to fill in the text part.

Check that both windows are drawn, that overlapping sections are handled correctly, and the text is written within each window. If this all works, the hardest part of the project is done. All that is required now is a few bells and whistles.

Writing text to a window

Having the capability to create and display windows is not of much use unless one can write text to them. This is the part covered next. Initially only line output will be considered. In other words the function will write only complete lines to the window, not one character at a time.

To write some text to a window, two items need to be specified. These are: which window and what text. A suitable prototype for this function would be:

```
wprint(struct window *, char *);
```

In other words the function requires the address of a structure defining which window to write into, and also the address of the character string to be written.

The process required is to copy the given text into the window text buffer starting at the current offset position. When the end of the text is reached fill to the end of the current line with blanks. Then update the offset position to indicate the start of the next line in the text buffer, and call the show function to display the text.

The above will work fine unless the text to be written exceeds the width of the window. If it does, then the text might need to be put partly at the end of the text buffer and partly at the start. In other words, the text may need to wrap round the end of the buffer. Simply check after saving each character whether the end of the buffer has been reached: if so, ensure the remaining characters get placed at the start.

To implement this function start by defining a local integer variable and preset its value to that of the text offset pointer. Then copy each character of the text string into the text buffer using the local variable as the array index. Increment this variable for each character and reset it to zero should it reach the end of the buffer. Once the text has been copied fill the remainder of the current line with blanks. The remainder operator (%) can be used to determine when the end of the current line has been reached. Then set the field offset variable to the value of the local variable, setting it back to zero if required. Finally call the show function described previously to write the text to the screen.

Test this by writing a number of lines to the top window, that is, the last one created. Write sufficient lines to cause the window text to scroll and check all works as expected. Try with short lines, long lines, and also with empty lines. When everything is working correctly write the test lines to a window other than the top–most one. This should highlight a problem which is covered in the next section but before reading on try to determine how this problem can be solved.

Screen ownership

As was probably noticed in the previous testing, there is a problem when writing text to a window that is partially or totally covered by another window. Clearly the text must be saved in the appropriate text buffer but some or all of it may not require displaying. The function that writes the characters to the screen needs to know to which window a particular character position belongs. If it is not the window being written to, then the character should not be displayed.

The simplest way to implement this requirement is to have an integer array the same size as the screen. Each element of the array represents one character position on the screen and contains a number. This is the number of the window which 'owns' this character position. Thus a character can only be written to the screen when the current window controls that character position.

To implement this proceed as follows. Write a function `winit` that requires no parameters. This function is to be called once to initialize the windowing system. The code which determines the screen address would logically be moved into this function. The other action required is to declare and initialize to zero, an array of type `int` that is the same size as the screen. This is going to be the screen ownership array. It is probably worth having a global integer pointer that is assigned sufficient memory using the function `malloc`.

The window create function then needs slight modification to set the window number variable in the structure. This should be set to one for the first window, two for the second, and so on. A good case for a static variable preset to one.

The function that displays the window border and name also requires slight modification. Here the area of the screen ownership array that matches the window being displayed (including the border) needs to be set to the current window number. This effectively claims that area of the screen.

Finally modify the window show routine, the one that writes the text to the display, so that characters are only written to screen positions where the screen ownership element matches the current window number.

Test these modifications by repeating the tests for the previous section. Writing lines to partially hidden windows should not now modify the covering windows.

Linked lists

We now have most of the functionality required to create and use text windows. The only item missing is the ability to change the order of the windows on the display; in other words, to raise an obscured, or partially obscured, window so that it is fully visible, or to move a visible one behind others. To do this some form of ordering is required. There are a number of different ways that the window order can be stored, the proposed linked list technique being only one.

Before writing code one must define what is required. Is the required order from front to back (i.e. most visible window first), or from back to front? The answer is probably both. The front to back ordering is required when creating a new window as this will need to be fully visible. While the back to front order would be useful when redrawing the screen as displaying the windows in that order makes the front window the last one written and hence fully visible.

If a refresher on linked lists is felt necessary refer to Chapter 28 which first raised this subject. Otherwise let's proceed to implement a doubly linked list.

First define two global pointers each of type `struct window *`. These are to be pointers to the top window and the bottom window and should be set to NULL in the initialization routine.

The window create function is the first to be modified. Each time a new window is created the 'next' pointer in the new window structure should be set to the value in the top pointer. Then, if the top pointer is not NULL, the 'previous' pointer in the window structure currently addressed by the top pointer should be set to the address of the new window structure. Otherwise, if the top pointer were NULL, then this is the first window and is therefore both the top and the bottom window. In this case set the bottom pointer to address this new structure. Finally the top pointer should then be set to the address of the new window structure. Thus whenever a new window is created it will be put at the top of the list. Also the one that was on top, if there was one, will now point backwards to the new structure. If unsure on this draw a series of diagrams similar to those used in Chapter 28 showing the actions as the first two windows are created.

In order to test the links have been set up correctly write a 'redraw' function that redraws the screen. It need not clear the screen, but should take the value in the bottom pointer and display that window. Then the backward pointer should be followed and the previous window displayed. This process should be repeated until the backward pointer is NULL. The display should be the same as for the previous output. If not, persevere!

Raising and lowering windows

Raising a window means bringing it to the front so that it is completely visible. Conversely, lowering a window involves moving it to the back. Both actions can be realized simply by changing the pointers around. As both raising and lowering a window first require the window to be removed from its current position in the lists, this seems a good contender for a function.

Start by writing a function that removes the given window from the list. The address of the window structure should be specified as a parameter to the function. The function should set the 'next' pointer in the previous window to the current windows' next pointer and the 'previous' pointer in the next window to the current previous pointer. Figure 36.4 shows the pointers before and after the middle window is removed.

Fig. 36.4   Removing a window from the linked lists

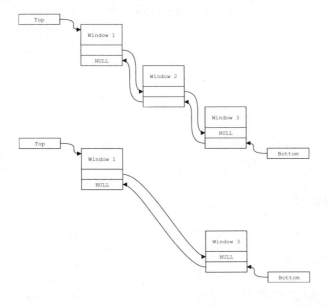

There is, as always, a slight snag. If either, or both, of the current window pointers are NULL then the `top` or `bottom` pointers will require changing. The code shown in Fig. 36.5 copes with the window being the first in the list (i.e. the top window). The variable p is the address of the window structure being removed.

Fig. 36.5   Removing a window (partial code)

```
if (p->prev == NULL)
   top = p->next;
else
   p->prev->next = p->next;
```

When happy you understand the given code, especially the double pointer reference, write the code required to handle the case where the window could be at the bottom of the list. Complete the removal function by setting both the pointers in the window structure being removed to NULL. This will simplify adding the window back into the list at a later time. Once written verify the function will correctly remove a given window when the original list contains one, two, or three windows.

Raising a window

Now back to the original topic of raising a given window. This function will require the address of the window structure to raise. Clearly if the given window is already the top window no action is required, and this should be checked for. Otherwise remove the window from the list using the previous function and then add the window to the top of the list. This mechanism was described earlier. Calling the redraw function will then update the display, hopefully with the requested window at the top.

Modify the test program so that two overlapping windows are drawn. After some keyboard input to action a delay, raise the first window. Then after another delay raise the second window. By the time this is working you should be in a good position to write a function to lower (i.e. move to the bottom) a given window. Try it and see.

Final testing

All the required functions have now been written: this is just a test program to demonstrate their use.

The program should create three partially overlapping windows. Then an infinite loop is made that writes the text 'This is line number xxx' to each window in turn. The standard function `sprintf` would be a good choice here. Obviously the line number is incremented each time round the loop. Write this much and test it, noting that ctrl C (i.e. the control key and letter C together) will terminate the program.

Next, some keyboard input is required within the loop but input should only be read once a key is pressed. The function kbhit (introduced in Chapter 20) will return a true value when a key has been pressed, and getch can be used to read the character. Add a call to kbhit and, if a true value is returned, then read the character using getch. If the value read was 27 decimal, which corresponds to the escape key, then terminate the loop. Now run the program and confirm that it can be terminated by pressing the escape key.

Finally let's use function keys F1–F3 to raise windows one to three, and the keys F4–F6 to lower them. If a function key in the range F1 to F10 is pressed then a call to getch will return a zero value. A subsequent call to getch will return a number in the range 59 to 68 that correspond to F1 to F10. Write a function that calls getch and returns the value read unless it is zero. If the first value read is zero, then issue a second call to getch and return a value in the range −1 to −10 that corresponds to keys F1–F10.

Now, within the loop in the test program, incorporate a switch statement to process a character when entered. This should check not only for the escape key as before but also for any of the function keys F1 to F6. Action the function keys by raising or lowering the required window.

Mouse interface

I know I said we wouldn't, but if there is a Microsoft–compatible mouse available let's try using it.

Accessing the mouse is done via the BIOS using the int86 function introduced in Chapter 29. The required mouse functions are: initialize, display, hide, and get attributes. Each of these functions is available through the int86 function using software interrupt 33 hex. Appendix H gives the details.

Start by writing a new test program that initializes the mouse using sub–function zero (i.e. register ax = 0). If the return status (in register ax) indicates a mouse is present then set some flag accordingly. If a mouse is available then use sub–function one (i.e. register ax = 1) to display the mouse cursor. At this point add a call to getchar to cause a pause and then hide the mouse cursor using sub function two. One point to note here is that if the show mouse cursor function is called twice, then two calls to the hide function will be required before it actually disappears.

Test this much and one should be able to move the cursor on the screen when the program is run. The mouse cursor should not be left on the screen once the program has finished.

When this much is working write a function to interrogate the mouse cursor position using sub–function three. The function should return the button status (from register bx) as its value. The return value will then be one if the left button is pressed, two for the right, three for both, and zero for no buttons pressed. The *x* and *y* co-ordinates should be returned in two parameters that will need to be passed as pointers, thus permitting values to be returned. A suitable function prototype would be:

```
int mget(int *, int *);
```

In order to return the co-ordinates as line and column numbers the values returned in the registers will require dividing by eight. Do this within the function.

To test this add a loop that repeatedly calls the `mget` function until a non-zero value is returned. Each time round the loop print out the *x* and *y* co-ordinates returned. This loop will require inserting in the test program where the mouse cursor is visible. Check the program works, and confirm that the row and column numbers returned are consistent with the mouse cursor position. Check particularly the four corners of the screen.

Now that the mouse interface is available, incorporate it into the window test program. Near the beginning of the program initialize the mouse interface and set a flag if the mouse is available. If there is no mouse then the flag should be false and no other mouse routines called. If a mouse is available then display the mouse cursor, and do not forget to hide it again at the end of the program. Within the loop of the main program then, if the mouse is available, status it and ignore it unless one of the two buttons are pressed. If either button is pressed then proceed as follows.

From the mouse co-ordinates determine in which window the mouse is by looking up the window number in the screen ownership map. Ignoring zero values (as these represent part of the screen without any window) then scan the list of windows from the top to find the address of the window structure with the given number. If the window structure address is found raise the window if the left button is pressed, and lower it for the right button.

This concludes the project but there is plenty of scope for using the functions developed. For example one could implement a top line menu system by defining a sequence of windows where the first line of each window was the menu name. These could be placed along the top of the screen and a larger window defined over the rest of the screen overwriting all but the top line of the menu windows. Selecting a menu name with the mouse and pressing button one would raise the menu so that it became fully visible. For this sort of application it might be worth having a way of preventing a window having a border. Anyway, there is room for experiment.

Windows project—getting started

```c
#include <stdio.h>

struct screen {
   unsigned char c;
   unsigned char a;
};

int nlines;                            /*   Number of lines    */
int ncols;                             /*   Number of columns  */
struct screen _far *sptr;              /*   Display address    */

int main(void);
void exit(int);
struct screen _far *getadr(void);

int main(void)
{
   int i;
   sptr = getadr();
   if (sptr == NULL)
      exit(1);
   for (i=0;  i < nlines*ncols;  i++)
      sptr[i].c = ' ';
   getchar();   exit(0);
   return 0;
}

struct screen _far *getadr(void)    /*   Get screen address   */
{
   unsigned char _far *p;
   nlines = 25;
   p = (unsigned char _far *) 0x00400049;
   switch (*p)                          /*   *p = screen mode   */{
      case 2:
      case 3:
      case 10:
             ncols = 80;
             break;
      case 0:
      case 1:
      case 9:
             ncols = 40;
             break;
      case 8:
             ncols = 20;
             break;
      case 7:
             ncols = 80;
             return((struct screen _far *) 0xB0000000);
      default:
             return(NULL);
   }
   return((struct screen _far *) 0xB8000000);
}
```

Windows project—creating a window

```c
#include <stdio.h>
#include <string.h>
#include <stdlib.h>

struct window {
   int height;
   int width;
   int xpos;
   int ypos;
   int num;
   int offset;
   unsigned char *text;
   unsigned char *name;
   struct window *prev;
   struct window *next;
};

int main(void);
void exit(int);
struct window *wcreate(int, int, int, int, char*);

int ncols = 80;
int nrows = 25;


int main(void)
{
   struct window *ptr;
   ptr = wcreate(5, 5, 30, 10, "Window 1");
   if (ptr != NULL)
     printf("Window structure created\n");
   else
     printf("Window creation failed\n");

   return 0;
}


struct window *wcreate(int xpos, int ypos, int width,
                  int height, char *name)
{
   int i;
   struct window *ptr;

/*   Check window will fit on screen   */

   if ((xpos < 1) || (xpos+width+1 >= ncols))
     return(NULL);
   if ((ypos < 1) || (ypos+height+1) >= nrows))
     return(NULL);

/*   Allocate window structure   */

   ptr = malloc(sizeof(struct window));
   if (ptr == NULL)
     return(NULL);
   ptr->height = height;
   ptr->width = width;
   ptr->xpos = xpos;
   ptr->ypos = ypos;
```

```
/*   Allocate and clear space for text   */

   ptr->text = malloc(height * width);
   if (ptr->text == NULL) {
     free(ptr);
     return(NULL);
   }
   for (i=0;  i<width*height;  i++)
     ptr->text[i] = ' ';

/*   Allocate space for window name   */

   ptr->name = malloc(strlen(name) + 1);
   if (ptr->name == NULL) {
     free(ptr->text);
     free(ptr);
     return(NULL);
   }
   strcpy(ptr->name, name);

/*   Preset remaining variables   */

   ptr->num = 0;
   ptr->offset = 0;
   ptr->prev = NULL;
   ptr->next = NULL;
   return(ptr);
}
```

Windows project—displaying windows

```
void wdisplay(struct window *ptr)              /*   Display window   */
{
   int i;
   unsigned char *c;
   struct screen _far *p;
   struct screen _far *q;

   p = sptr + ((ptr->ypos - 1) * ncols + ptr->xpos - 1);
   q = p + (ptr->height + 1 ) * ncols;

/*   Draw top and bottom lines   */

   for (i=1; i <= ptr->width; i++)
      p[i].c = q[i].c = 205;

/*   Draw four corners   */

   p[0].c = 201;
   p[i].c = 187;
   q[0].c = 200;
   q[i].c = 188;

/*   Write in the window name   */

   c = ptr->name;
   i = (ptr->width - strlen(c)) / 2;
   p[i-1].c = 181;
   while (*c != '\0')
      p[i++].c = *(c++);
   p[i].c = 198;

/*   Draw the sides   */

   for (i=0; i<ptr->height; i++) {
      p += ncols;
      p[0].c = p[ptr->width+1].c = 186;
   }
}
```

Windows project—displaying window text

```
void wshow(struct window *ptr)        /*   Display text   */
{
   int i, j, off;
   unsigned char *c;
   struct screen _far *p;

   p = sptr + (ptr->ypos * ncols + ptr->xpos);
   c = ptr->text;                               /*   Text address   */
   off = ptr->offset;                           /*   Starting offset */

   for (i=0; i<ptr->height; i++) {
      for (j=0; j<ptr->width; j++)
         p[j].c = c[off++];
      p += ncols;
      if (off >= ptr->width*ptr->height)
         off = 0;
   }
}
```

Windows project—writing text to a window

```
void wprint(struct window *ptr, char *text)
{
   int  i = ptr->offset;
   int  limit = ptr->width * ptr->height;

   while (*text != '\0') {            /*  Copy text              */
      ptr->text[i++] = *(text++);
      if (i >= limit)                 /*  Check for wrap         */
         i = 0;
   }

   while ((i % ptr->width) != 0)      /*  Blank fill line        */
      ptr->text[i++] = ' ';

   if (i >= limit)                    /*  Check if last line     */
      i = 0;

   ptr->offset = i;                   /*  Update start offset    */
   wshow(ptr);                        /*  Display the text       */
}
```

Windows project—storing screen ownership

```
void winit(void)                          /*  Initialize system  */
{
    int i;
    sptr = getadr();                      /*  Get screen address  */
    if (sptr == NULL)
        exit(1);

    for (i=0; i < ncols*nrows; i++) {     /*  Blank fill screen  */
        sptr[i].c = ' ';
        sptr[i].a = 0x07;
    }

    i = ncols * nrows * sizeof(int);
    optr = malloc(i);                     /*  Assign map space  */
    if (optr == NULL)
        exit(1);

    for (i=0; i < ncols*nrows; i++)       /*  Zero fill  */
        optr[i] = 0;
}




void wdisplay(struct window *ptr)
{
    int i, j;
    unsigned char *c;
    int *m;
    struct screen _far *p;
    struct screen _far *q;
/*  Flag window ownership  */

    m = optr + ((ptr->ypos -1 ) * ncols + ptr->xpos - 1);
    for (i=0; i <= ptr->height + 1; i++) {
        for (j=0; j <= ptr->width + 1; j++)
            m[j] = ptr->num;
        m += ncols;
    }

/*  Draw border  */

    p = sptr + ((ptr->ypos - 1) * ncols + ptr->xpos - 1);
    q = p + (ptr->height + 1 ) * ncols;

/*  Draw top and bottom lines  */

    for (i=1; i <= ptr->width; i++)
        p[i].c = q[i].c = 205;

/*  Draw four corners  */

    p[0].c = 201;
    p[i].c = 187;
    q[0].c = 200;
    q[i].c = 188;
```

```
/*   Write in the window name   */

   c = ptr->name;
   i = (ptr->width - strlen(c)) / 2;
   if (i > 0) {
     p[i-1].c = 181;
     while (*c != '\0')
       p[i++].c = *(c++);
     p[i].c = 198;
   }

/*   Draw the sides   */

   for (i=0; i < ptr->height; i++) {
     p += ncols;
     p[0].c = p[ptr->width+1].c = 186;
   }

/*   Write text   */

   wshow(ptr);
}


void wshow(struct window *ptr)
{
   int i, j, off;
   int *m;
   unsigned char *c;
   struct screen _far *p;

   p = sptr + (ptr->ypos * ncols + ptr->xpos);
   m = optr + (ptr->ypos * ncols + ptr->xpos);
   c = ptr->text;
   off = ptr->offset;
   for (i=0; i<ptr->height; i++) {           /*   For each line   */
     for (j=0; j<ptr->width; j++) {
       if (m[j] == ptr->num)                 /*   Check owner of screen
         p[j].c = c[off];
       off++;
     }
     p += ncols;
     m += ncols;
     if (off >= ptr->width * ptr->height)
       off = 0;
   }
}
```

Windows project—raising and lowering windows

```
void wremove(struct window *p)        /*  Remove window from lists  */
{
   if (p->prev == NULL)
      top = p->next;
   else
      p->prev->next = p->next;

   if (p->next == NULL)
      bottom = p->prev;
   else
      p->next->prev = p->prev;

   p->next = p->prev = NULL;
}


void wraise(struct window *p)         /*  Raise given window  */
{
   if (p != top) {
      wremove(p);
      p->next = top;
      if (top != NULL)
         top->prev = p;
      top = p;
      wredraw();
   }
}


void wlower(struct window *p)         /*  Lower given window  */
{
   if (p != bottom) {
      wremove(p);
      p->prev = bottom;
      if (bottom != NULL)
         bottom->next = p;
      bottom = p;
      wredraw();
   }
}
```

Windows project—main test program

```
#include <stdio.h>
#include <conio.h>
#include <dos.h>

int main(void);
void exit(int);
int mgetch(void);


int main(void)
{
   char line[40];
   int c, n;
   int num = 0;
   int row, col;
   int flag = TRUE;
   union REGS regs;
   struct window *p, *ptr1, *ptr2, *ptr3;

   winit();

   ptr1 = wcreate(5, 5, 30, 10, "Window 1");
   if (ptr1 == NULL)
      exit(2);
   wdisplay(ptr1);

   ptr2 = wcreate(12, 9, 30, 10, "Window 2");
   if (ptr2 == NULL)
      exit(2);
   wdisplay(ptr2);

   ptr3 = wcreate(30, 12, 40, 11, "Window 3");
   if (ptr3 == NULL)
      exit(2);
   wdisplay(ptr3);

   while (flag) {
      sprintf(line, "This is line number %d", num++);
      wprint(ptr1, line);
      wprint(ptr2, line);
      wprint(ptr3, line);

      if (kbhit()) {              /*  Test for key press  */
         switch (mgetch()) {
            case -1:              /*  Raise window one    */
                     wraise(ptr1);
                     break;
            case -2:              /*  Raise window two    */
                     wraise(ptr2);
                     break;
            case -3:              /*  Raise window three  */
                     wraise(ptr3);
                     break;
            case -4:              /*  Lower window one    */
                     wlower(ptr1);
                     break;
            case -5:              /*  Lower window two    */
                     wlower(ptr2);
                     break;
            case -6:              /*  Lower window three  */
                     wlower(ptr3);
                     break;
```

```
            case 27:                    /*   Escape key pressed   */
                        flag = FALSE;
                        break;
            }
        }
    }
    return 0;
}



int mgetch(void)                /*  Get extended character   */
{
    int c;
    c = getch();
    if (c == 0)                 /*   If function key   */
        c = 58 - getch();
    return c;
}
```

Windows project—mouse test program

```
#include <stdio.h>
#include <dos.h>

#define TRUE 1
#define FALSE 0

int main(void);
void exit(int);
int mstatus(int *, int *);

int main(void)
{
  int mflag;
  int row, col;
  union REGS regs;
  regs.x.ax = 0;                    /*  Check if mouse available  */
  int86(0x33, &regs, &regs);
  mflag = (int) regs.x.ax;
  if (!mflag) {
    fprintf(stderr, "Mouse driver not installed\n");
    exit(1);
  }

  regs.x.ax = 1;                    /*  Show mouse cursor  */
  int86(0x33, &regs, &regs);

  while (mstatus(&row, &col) == 0)
    printf("%d,%d    \r", row, col);

  regs.x.ax = 2;                    /*  Hide mouse cursor  */
  int86(0x33, &regs, &regs);
  return 0;
}


int mstatus(int *row, int *col)    /*  Get mouse status  */
{
  union REGS regs;
  regs.x.ax = 3;
  int86(0x33, &regs, &regs);
  *row = (int) regs.x.dx / 8;
  *col = (int) regs.x.cx / 8;
  return (int) regs.x.bx;
}
```

Windows project—complete solution

```c
#include <stdio.h>
#include <string.h>
#include <stdlib.h>
#include <conio.h>
#include <dos.h>

#define TRUE 1
#define FALSE 0

struct window {
   int height;
   int width;
   int xpos;
   int ypos;
   int num;
   int offset;
   unsigned char *text;
   unsigned char *name;
   struct window *prev;
   struct window *next;
};

struct screen {
   unsigned char c;
   unsigned char a;
};

struct screen _far *sptr;    /*   Address of screen        */
struct window *top;          /*   Address of top window    */
struct window *bottom;       /*   Address of last window   */
int *optr;                   /*   Screen ownership map      */
int mflag;                   /*   Mouse available flag      */
int ncols;                   /*   Number of screen columns */
int nrows;                   /*   Number of screen rows     */

int main(void);
void exit(int);
void winit(void);
struct window *wcreate(int, int, int, int, char*);
void wdisplay(struct window *);
void wshow(struct window *);
void wprint(struct window *, char *);
void wredraw(void);
void wremove(struct window *);
void wraise(struct window *);
void wlower(struct window *);
int mgetch(void);
int mstatus(int *, int *);


int main(void)
{
   char line[40];
   int c, n;
   int num = 0;
   int row, col;
   int flag = TRUE;
   union REGS regs;
   struct window *p, *ptr1, *ptr2, *ptr3;

   winit();
```

```
ptr1 = wcreate(5, 5, 30, 10, "Window 1");
if (ptr1 == NULL)
   exit(2);
wdisplay(ptr1);

ptr2 = wcreate(12, 9, 30, 10, "Window 2");
if (ptr2 == NULL)
   exit(2);
wdisplay(ptr2);

ptr3 = wcreate(30, 12, 40, 11, "Window 3");
if (ptr3 == NULL)
   exit(2);
wdisplay(ptr3);

while (flag) {
   sprintf(line, "This is line number %d", num++);
   wprint(ptr1, line);
   wprint(ptr2, line);
   wprint(ptr3, line);

   if (mflag) {                      /*  If mouse available  */
      c = mstatus(&row, &col);
      if (c != 0) {
         n = *(optr + row * ncols + col);/* window number  */
         p = top;
         while (p != NULL) {        /*  Get window address  */
            if (p->num == n)
               break;
            p = p->next;
         }
         if (p != NULL) {
            if (c & 1)
               wraise(p);
            if (c & 2)
               wlower(p);
         }
      }
   }

   if (kbhit()) {                    /*  Test for key press  */
      switch (mgetch()) {
         case -1:                    /*  Raise window one    */
               wraise(ptr1);
               break;
         case -2:                    /*  Raise window two    */
               wraise(ptr2);
               break;
         case -3:                    /*  Raise window three  */
               wraise(ptr3);
               break;
         case -4:                    /*  Lower window one    */
               wlower(ptr1);
               break;
         case -5:                    /*  Lower window two    */
               wlower(ptr2);
               break;
         case -6:                    /*  Lower window three  */
               wlower(ptr3);
               break;
         case 27:                    /*  Escape key pressed  */
               flag = FALSE;
               break;
      }
   }
}
```

```
if (mflag) {
  regs.x.ax = 2;                        /*  Hide mouse cursor    */
  int86(0x33, &regs, &regs);
}
return 0;
}


struct window *wcreate(int xpos, int ypos, int width,
                       int height, char *name)
{
  int i;
  struct window *ptr;
  static int wnum = 1;

/*  Allocate window structure  */

  ptr = malloc(sizeof(struct window));
  if (ptr == NULL)
    return(NULL);
  ptr->height = height;
  ptr->width = width;
  ptr->xpos = xpos;
  ptr->ypos = ypos;

/*  Allocate and clear space for text  */

  ptr->text = malloc(height * width);
  if (ptr->text == NULL) {
    free(ptr);
    return(NULL);
  }
  for (i=0; i<width*height; i++)
    ptr->text[i] = ' ';

/*  Allocate space for window name  */

  ptr->name = malloc(strlen(name) + 1);
  if (ptr->name == NULL) {
    free(ptr->text);
    free(ptr);
    return(NULL);
  }
  strcpy(ptr->name, name);

  ptr->num = wnum++;
  ptr->offset = 0;

/*  Add new window to linked list  */

  if (top != NULL)
    top->prev = ptr;
  else
    bottom = ptr;
  ptr->next = top;
  ptr->prev = NULL;
  top = ptr;
  return(ptr);
}

void wdisplay(struct window *ptr)
{
  int i, j;
  unsigned char *c;
  int *m;
  struct screen _far *p;
  struct screen _far *q;
```

```
/*   Flag window ownership   */

   m = optr + ((ptr->ypos - 1) * ncols + ptr->xpos - 1);
   for (i=0; i < ptr->height+2; i++) {
      for (j=0; j < ptr->width+2; j++)
         m[j] = ptr->num;
      m += ncols;
   }

/*   Draw border   */

   p = sptr + ((ptr->ypos - 1) * ncols + ptr->xpos - 1);
   q = p + (ptr->height + 1 ) * ncols;

/*   Draw top and bottom lines   */

   for (i=1; i <= ptr->width; i++)
      p[i].c = q[i].c = 205;

/*   Draw four corners   */

   p[0].c = 201;
   p[i].c = 187;
   q[0].c = 200;
   q[i].c = 188;

/*   Write in the window name   */

   c = ptr->name;
   i = (ptr->width - strlen(c)) / 2;
   p[i-1].c = 181;
   while (*c != '\0')
      p[i++].c = *(c++);
   p[i].c = 198;

/*   Draw the sides   */

   for (i=0; i<ptr->height; i++) {
      p += ncols;
      p[0].c = p[ptr->width+1].c = 186;
   }

/*   Write text   */

   wshow(ptr);
}


void wshow(struct window *ptr)
{
   int i, j, off;
   int *m;
   unsigned char *c;
   struct screen _far *p;

   p = sptr + (ptr->ypos * ncols + ptr->xpos);
   m = optr + (ptr->ypos * ncols + ptr->xpos);
   c = ptr->text;                            /*   Text address     */
   off = ptr->offset;                        /*   Starting offset  */

   for (i=0; i<ptr->height; i++) {           /*   For each line    */
      for (j=0; j<ptr->width; j++) {         /*   For each column  */
         if (m[j] == ptr->num)
            p[j].c = c[off];                 /*   Display character */
         off++;
      }
```

```
      p += ncols;                           /*  Move  on  one  line   */
      m += ncols;
      if (off >= ptr->width*ptr->height)    /*  Check for wrap   */
        off = 0;
   }
}


void wprint(struct window *ptr, char *text)
{
   int i = ptr->offset;
   int limit = ptr->width * ptr->height;

   while (*text != '\0') {            /*  Copy text into buffer */
      ptr->text[i++] = *(text++);
      if (i >= limit)                 /*  Check for wrap around */
        i = 0;
   }

   while ((i % ptr->width) != 0)   /*  Blank fill line         */
      ptr->text[i++] = ' ';

   if (i >= limit)                   /*  Check if last line      */
      i = 0;

   ptr->offset = i;                  /*  Update start of window */
   wshow(ptr);                       /*  Display the text        */
}


void winit(void)                     /*  Initialize system       */
{
   int i;
   sptr = getadr();                  /*  Get screen address      */
   if (sptr == NULL)
      exit(1);

   for (i=0; i<ncols*nrows; i++) { /*  Blank fill screen        */
      sptr[i].c = ' ';
      sptr[i].a = 0x07;
   }

   i = nrows * ncols * sizeof(int);
   optr = malloc(i);                             /*  Assign map space   */
   if (optr == NULL)
      exit(1);

   for (i=0; i < ncols*nrows; i++)       /*  Zero fill   */
      optr[i] = 0;

    regs.x.ax = 0;                         /*  Initialize mouse   */
   int86(0x33, &regs, &regs);
   mflag = (int) regs.x.ax;
   if (!mflag)
      return;

   regs.x.ax = 1;                         /*  Display cursor   */
   int86(0x33, &regs, &regs);
}


void wredraw(void)                          /*  Redraw screens   */
{
   struct window *ptr = bottom;
   while (ptr != NULL) {
      wdisplay(ptr);
      ptr = ptr->prev;
```

```
   }
}
void wremove(struct window *p)       /*   Remove given window   */
{
   if (p->prev == NULL)
     top = p->next;
   else
     p->prev->next = p->next;

   if (p->next == NULL)
     bottom = p->prev;
   else
     p->next->prev = p->prev;

   p->next = p->prev = NULL;
}


void wraise(struct window *p)       /*   Raise given window   */
{
   if (p != top) {
     wremove(p);
     p->next = top;
     if (top != NULL)
       top->prev = p;
     top = p;
     wredraw();
   }
}


void wlower(struct window *p)       /*   Lower given window   */
{
   if (p != bottom) {
     wremove(p);
     p->prev = bottom;
     if (bottom != NULL)
       bottom->next = p;
     bottom = p;
     wredraw();
   }
}


int mgetch(void)                /*   Get extended character   */
{
   int c;
   c = getch();
   if (c == 0)
     c = 58 - getch();
   return c;
}


int mstatus(int *row, int *col)   /*   Get mouse status   */
{
   union REGS regs;
   regs.x.ax = 3;
   int86(0x33, &regs, &regs);
   *row = (int) regs.x.dx / 8;
   *col = (int) regs.x.cx / 8;
   return (int) regs.x.bx;
}
```

37

Satellite image processing

Uses

This project is the big one: successful completion will demonstrate a good grasp of the C language and its use. As it is such a large project it should not be tackled first. Practise on some of the earlier projects even if they are not found to be so interesting.

In order to display the image an EGA or VGA display is required. To print the image an Epson (or compatible) dot matrix printer is required.

Overview

The idea of this project is to write a suite of programs that will enable a satellite image to be displayed on the screen, and also printed on paper. On the way to this objective, drawing and printing a map of the major continents will also be attempted. Note that several programs will be required, not one monolithic one. Recognizing that several smaller programs can often act better than one large one, is one of the harder lessons to learn. All too often the temptation is to expand an existing program to add a new feature. As the required features expand, the program becomes less and less easy to modify, and the debugging time increases. The art is to design small programs that do specific jobs, and do them well, then use these programs together to achieve the desired effect. This will be demonstrated as the project progresses.

Displaying graphics

Both the EGA and VGA displays have two modes of working. One is text mode, the other is graphical mode. In text mode the smallest unit definable on the screen is one character shape. This shape is usually one from a predefined range but can be defined by the programmer. By contrast, the graphical mode of operation allows each point, or pixel, to be defined independently. For this reason it is sometimes referred to as all points addressable (or APA) mode. Both text and APA modes have several sub–modes. For example text modes can be 40 or 80 characters wide, 25 or 30 lines, colour or black and white.

Similarly the number of pixels in graphical modes varies as does the number of colours available for each pixel. Appendix C lists the basic screen modes.

The display controller can also handle multiple pages, where one page is a complete displayable image. Only one page can be displayed at a time but the controller can switch pages on demand. This technique can be used to produce crude animation. The number of pages varies with screen mode and the amount of display controller memory.

For this project screen mode sixteen (10 hex) will be assumed as this is available on both EGA and VGA displays. Also only one page (page zero) will be used. This is because not all EGA cards have sufficient memory to support more than one page in this mode.

Changing screen modes

If a program changes the screen mode it is a good idea to restore the original mode before the program terminates. If this is not done then subsequent screen output may not be displayed correctly. It is important to catch all possible exits from the program.

To set the screen mode requires a BIOS request. This was described in Chapter 29 but, before setting the screen mode, the original mode must be determined so that this can be restored at the end of the program. The first exercise in Chapter 29 involved writing a program to determine the screen mode. Using this code, together with code to set the screen mode as shown in Fig. 29.1, write a function that sets the screen mode to a given value, and returns the original mode. The function should take one integer parameter, the required screen mode and the function's return value should be the original screen mode as an integer value. Incorporate this function into a program that should set the screen mode to sixteen (10 hex), wait for a character to be entered and then restore the original screen mode. When run the screen should be cleared twice, once each time the screen mode is changed.

Drawing pixels

Drawing pixels can be as hard or as easy as one likes. Generally the easier the code, the slower it executes. Initially go for the slowest possible, the BIOS routines. BIOS software interrupt 10 hex, sub-function twelve (0C hex) sets a single pixel to a given value. Assuming screen mode sixteen is in effect then the valid pixel values are zero to fifteen and correspond to sixteen different colours. The screen mode also determines the number of pixels on the screen. For mode sixteen there are 640 pixels across by 350 down, the top left pixel being at position (0,0). The BIOS functions assume that all values passed to them are valid, you have been warned. So the next function to write will draw a single point on the screen. It should take three integer parameters. These are the horizontal position, vertical position, and the required pixel value. These are passed to the BIOS function as follows:

```
register  ah  =  0x0C
register  al  =  pixel value  (0-15)
register  bh  =  Page number  (0)
register  cx  =  X co-ordinate  (0-639)
register  dx  =  Y co-ordinate  (0-349)
```

Write this function but keep the code as small as possible as it will be used frequently. To fill the entire screen would take 224 000 calls, thus do not bother to check that the values given are within range.

When testing this function, two tips. Firstly, avoid using pixel value zero. This corresponds to black and makes it rather difficult to see. Secondly, don't write a single point, write a cluster or a line as these will then be distinguishable from dirt on the screen.

Drawing an outline globe

Having the ability to draw a point now enables more ambitious projects to be attempted. The next goal is to draw an outline map of the globe similar to that shown in Fig. 37.1. This is quite a large step so proceed carefully and check each stage as it is written.

Fig. 37.1 Global map drawn using points

Start by drawing a circle. To do this plot a series of points where the screen co-ordinates *x* and *y* are given by the formulae:

```
x = xc + r × sin(angle)
y = yc - r × cos(angle)
```

The angle is in the range 0 – 360°, r is the radius, and the point (x_c, y_c) is the centre of the circle. The trigonometric functions `sin()` and `cos()` both require a parameter of type `double`, which is in radians. To convert degrees to radians multiply by π/180 (i.e. 0.0175). The trigonometric functions will return a value of type `double` which will be in the range −1.0 to +1.0. Suitable values for the other parameters are:

```
r  = 150.0
xc = 319.0
yc = 174.0
```

The co-ordinates of the point on the screen will need to be integer but do all the calculations using variables of type `double`, and only cast to `int` at the last moment. That is enough theory for the moment. Try writing a program to draw a circle. Remember to use the include file `math.h` and do not forget to set graphics mode and restore the original screen mode after waiting for input. Draw the circle in any colour (or several colours) but not black!

Do not be surprised that the circle is a bit squashed. This is because the screen resolution in the horizontal and vertical directions differ. Determine a scale factor to multiply the horizontal co-ordinate by, to obtain a true circle. If s is the required horizontal scale factor then the *x* co-ordinate calculation becomes:

```
x = xc + r × s × sin(angle)
```

Next replace the circle drawing section with code to draw the lines of latitude and longitude, start with the lines of longitude (i.e. the North–South lines). For longitude values of −90°, −60°, −30°, 0°, 30°, 60° and 90° plot the points with latitude values in the range of −90° to +90° in steps of one degree. The screen co-ordinates can be derived using the formulae given in Fig. 37.2. Do not forget to convert the degree value to radians and also to scale the horizontal co-ordinate to get the correct aspect ratio.

Fig. 37.2 Co-ordinate mapping formulae

```
x = xc + r × cos(latitude) × sin(longitude)
y = yc - r × sin(latitude)
```

Having obtained the lines of longitude add the lines of latitude. Use the same formulae but this time with the latitude taking values of −60°, −30°, 0°, 30°, and 60° and the longitude going from −90° to +90° in steps of one degree.

After the grid lines comes the detail. A data file is supplied that contains the latitude and longitude co-ordinates of the major coastlines. The next stage is to read these points and plot them on the screen. This involves two parts, one to read the data, and the other to convert the data to the screen co-ordinates.

Reading the data has been made more interesting by supplying the data in a binary format. This makes the data smaller but slightly harder to read. It also makes the data less portable between different hardware and would thus not normally be recommended.

Each co-ordinate is stored as four bytes in the data file; the first two bytes being the latitude and the second two the longitude. Each pair of bytes form an integer which, after division by 100.0, gives the required angle in degrees. A co-ordinate pair can be read using the following code:

```
int c[2];
double latitude, longitude;
if (fread(c, sizeof(int), 2, fh) == 2) {
   latitude = (double) c[0] / 100.0;
   longitude = (double) c[1] / 100.0;
}
```

Write a function that reads co-ordinate pairs, converts the integer values from degrees to radians, determines the x and y co-ordinates on the screen (using the formulae given in Fig. 37.2), and plots the points on the screen. One thing to remember is that if the longitude is greater than +90°, or less than −90°, then the point is on the far side of the globe and need not be plotted. When successful the output should be as shown in Fig. 37.1.

As a bit of fun try adding 90° to the longitude value of every co-ordinate after the value has been read from the file. This will have the effect of rotating the drawn image. To rotate the image the other way subtract 90°. It is left as an exercise to draw the globe rotated by 180°.

Printing the screen

For this section an Epson dot matrix printer, or compatible printer, is required. If this is not available other printers may be substituted but they must be able to print bit image graphics and have the appropriate manual available to say how to do this.

The two new items of information required here are the ability to read the value of a particular pixel on the screen and how to send these values to the printer. The first is the easier so this is where to start.

Just as there is a BIOS request to write a given pixel to the screen so there is a corresponding read facility. The speed of access to a given cell is not very impressive: reading every pixel on the screen could well take over a minute.

The BIOS read pixel request is identical to that required to write a pixel except that the function code is 0x0D and the pixel value is returned in register `al` rather than specified by it. Thus the required input to the BIOS interrupt 10 hex is:

```
register  ah  =  0x0D
register  bh  =  Page number  (0)
register  cx  =  x co-ordinate  (0-639)
register  dx  =  y co-ordinate  (0-349)
```

Write a function which given an *x* and *y* screen co-ordinate returns the pixel value. Unless confident, test this out with a simple program that sets a pixel to a value and then reads it back. The value read should obviously be the same as that written.

That was the easy bit: now comes the printer output. Epson, and compatible printers, have a print head that can generate a vertical line of dots. In graphical mode up to eight dots are produced for each byte of data. The dots are separated by one seventy second of an inch, so if all eight pins are fired then a vertical line one ninth of an inch high will be produced. Each data byte sent to the printer is interpreted so that if a bit in the data byte is set (i.e. a one) then the corresponding pin produces a dot. The most significant bit in the data byte corresponding to the top most pin. Figure 37.3 shows this mapping.

The overall strategy for the printing function is to initialize the printer, then scan across the screen in bands eight pixels high and transmit the data to the printer. After 43 bands the end of the screen will be reached and any tidying up can be done.

Fig. 37.3   Mapping printer pins to the data byte

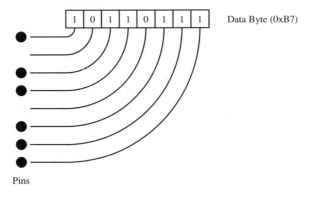

Pins

Initializing the printer is required mainly to set the amount by which the paper is advanced after each line. Obviously it must be equal to the height of the eight dots printed but this is not the default. To set the line spacing an escape sequence must be sent to the printer. An escape sequence is a group of bytes that are not printed, but are processed internally by the printer to change the way it operates. Escape sequences start with the ASCII character ESC (hence the name) and are usually followed by an alphabetic character. More data bytes may be required to complete the escape sequence. For example the sequence ESC @ will initialize the printer while ESC E sets bold or emphasized mode. The ASCII character ESC has a decimal value of 27 (33 octal).

The escape sequence to set the line spacing is ESC A which requires a numeric value in the next byte that is the number of seventy seconds of an inch between adjacent lines. To set the line spacing to the required one ninth of an inch (i.e. eight seventy seconds) would require:

ESC A <8>   (Where <8> means the numeric value 8)

These codes can be written using:

```
printf("\033A\010");
```

Then for each horizontal line (or band) of data, the printer has to be told that it has to use bit image or graphical mode rather than its normal character mode. This requires the escape sequence:

ESC K <xx> <yy> data

The two bytes <xx> and <yy> determine the number of data bytes that follow. The number of data bytes is given by yy * 256 + xx. The number of dots printable accross the paper is 480 (816 if using a wide carriage printer). Following the data a new line character (\n) is required to advance the paper ready for the next line. A form feed character (\f) after the last band of data would then advance the paper to the top of the next page.

Writing the data to the printer requires binary mode output. This is because the bit map graphics data could contain new line codes which must not be translated into the two byte form used by DOS. On many systems the following statement can be used:

```
fh = fopen("LPT1", "wb");
```

If this is accepted then any output will be sent either to the printer or to a file called 'LPT1'. If on testing a file is produced, try using the name 'LPT1:' instead. It would be nice if one of these two methods sent output directly to the printer, but this is not critical. If the output is written to a file then this can always be sent to the printer using the command:

```
copy filename LPT1:
```

We now have all the information required to dump the current screen contents in a format suitable for printing. Start by writing a function that opens LPT1 in binary write mode, checks that the open was successful, and then closes it. As the function will be called while a graphics screen is displayed writing error messages is not recommended. So if there is any error, simply exit the function with a false value. A true return value will indicate that the dump was successful.

Next add the section to initialize the printer. This will involve setting the vertical line spacing as described earlier and also, possibly, clearing the automatic end of page skip. Often by default the printer automatically advances the paper by two or three lines when the bottom of the page is reached to prevent text being written on the paper perforations. To cancel this automatic paper advance send the escape sequence:

```
ESC O
```

Finally the interesting bit, actually printing the screen contents. To dump the full screen, the print head will need to cross the paper 43 times. On each pass a band eight dots high will be produced. Simple arithmetic gives a total of 344 dots which is slightly less than the 350 dots on the screen. Just for the moment we will ignore the last six rows of the screen. As the screen is 640 pixels accross and the printer can only print 480 dots per line, some horizontal scaling will be required. Probably the easiest is to take each pair of screen pixels and print a single dot if either (or both) are non-zero.

To print each band, first output the escape sequence to set the printer into bit image mode (ESC K <xx> <yy>), setting the data byte count to 320 (the number of dots across the page). Then write the required 320 data bytes, each byte being built from the next pair of eight vertical screen pixels, starting from the left–hand edge. When the screen pixel has a zero value, set the corresponding bit in the data byte to zero. Any other screen pixel value should set the bit to a one. The following code could be useful, but is not mandatory:

```
byte = (byte << 1) | (pixel ? 1 : 0);
```

After the 320 data bytes output a new line character to advance the paper. Also, probably only after all the above code is working, add a form feed code (\f) after the last line to move the paper to the top of the next page.

All that remains is to test it, which is not trivial. Almost certainly the first attempt will fail and probably hang the printer. If a solution to the file dump project is available, then this could be used to verify that the output looks reasonable if the output is temporarily written to a file rather than directly to the printer. A bit of decoding of the bit image data with a pencil and paper may not be wasted. Otherwise just persevere, and remember to reset the printer before each attempt as, if given incorrect data, the printer will often hang.

Drawing images

In this text an image is defined as being a graphical picture composed of areas of colour rather than by lines. As an example, a photograph would be an image while a picture drawn using pen and ink would not. Despite the fact then the pen and ink drawing may give the effect of shading, it is still constructed from solid lines. Figure 37.4 shows the output required for the next section.

Fig. 37.4   Satellite image

Drawing rasters

When drawing images it is normal to calculate and then draw one horizontal row of pixels (known as a raster) at a time. Therefore a function that takes an *x* and *y* co-ordinate together with an array of pixel values and displays one horizontal raster seems to be a good idea. Write such a function making use of the pixel plotting function written earlier. Four parameters are probably required. Two are the *x* and *y* co-ordinates of the start of the line (assume this is the left most pixel). Another parameter would be the address of the array containing the pixel values, while the last parameter would specify the number of pixel values given. The pixel array could be of type `char` as the maximum value will be 15.

Test the function by filling the display with 16 colour bars, each bar being 40 pixels wide. This is probably most easily accomplished by filling a 640 element character array with the numbers 0 to 15 and then calling the raster plotting routine 350 times, altering the *y* co-ordinate on each call.

Satellite data format

The data sent by the satellite is a sequence of unsigned eight–bit values. Each value is therefore in the range 0 to 255. These represent intensities measured by the onboard sensors and can be visible light readings, infra–red or water saturation. The sensors are not defined as having a linear, or even regular, response throughout their range. The data originate from one of the Meteosat range of satellites. These scan the entire globe in one image. The original image is 2500 by 2500 points (6.25 million pixels). In fact three such images are sent on each scan, one sensitive to visible light, one to infra–red, and the last for water vapour concentration. This project only involves the visible image as this demonstrates all the required features.

Getting started

Whatever form the supplied data is in, the next step must be to decide on a general data file format. This is because the next few stages of the project require a number of small programs, each of which manipulate the image data in some way. Having a single data format that is read and written by each of these programs enables them to be used in different combinations. Chaining programs together like this can often lead to a number of simple programs performing better than a single large one.

For an image data file the most significant information is its size. Without this the data cannot be correctly interpreted. One other useful item would be the resolution of the data. The original has values in the range 0 to 255, but the display can only handle 16 values (256 if using a VGA at a lower resolution). It is therefore suggested that each file starts with a six byte header. This header is defined as follows:

```
byte 1 * 256 + byte 2 =   Image width
byte 3 * 256 + byte 4 =   Image height
byte 5 * 256 + byte 6 =   Data resolution
```

In every case a byte is taken as an unsigned value in the range 0 to 255. The resolution gives the number of colours (or intensities) in the data. For the original satellite data this will be 256.

The actual size of the supplied image will vary with its source. If the data have been obtained via the internet, then the size will be 2500 by 2500. If obtained on disk then examine the README.TXT file on the same disk for the image size.

The first program required needs to read the supplied data and re-write it but adding a six byte header at the start. The reading and writing must be done in binary mode and the generated file should be six bytes longer than the original. If the file dumping project has been done then this could be used to verify the created file looks about right, do not try to check every byte!

Very little help is going to be given here; after all, Chapter 21 gives a suitable solution as one of the examples. Minor modifications will be required to add the six–byte header and to use binary mode when opening the files. Also replacing the single byte read and write with block read and write (i.e. `fread` and `fwrite`) would improve the performance. The file names could be specified as parameters, as could the image size and resolution, but initially they can be hard coded. All right, just a bit more help: if the image size was 1024 by 836 pixels, and the resolution was 256 colours, then a suitable header could be written using:

```
fputc(1024/256, fout);
fputc(1024%256, fout);
fputc(836/256, fout);
fputc(836%256, fout);
fputc(256/256, fout);
fputc(256%256, fout);
```

Write a suitable program and verify that the generated data file is indeed six bytes longer than the original. If possible check part of the file contents to check all is well.

Image scaling

The next program required is to compress or scale one image into a more usable size. For example, an image 1024 by 836 will not fit on the display which is 640 by 350 dots. Thus some form of compression is required. It is suggested that only integer compression is attempted. In other words, for the given example, every two pixels horizontally and every three vertically could be averaged to form one output pixel. This would result in an image 512 by 276, extra pixels being ignored.

Write a program that reads the standard data file format, as defined in the previous section, and also requires two numbers which are the number of cells to average in the horizontal and vertical directions. These numbers can be entered either as program parameters or interactively.

Start by writing the main function that should read the source file header and print out the original image size. Check this much works. Then add the code to enable the compression factors to be entered. To check that this much works, print out the new image size. If these are as expected write out a suitable six byte header to the output file. Do not forget to use binary mode for all data files. Again, check this much works before progressing.

The next stage is to do the compression. A common failing is to assume that the entire image data must be read in, compressed, and then written out. This is very inefficient and on many systems would not be possible due to a lack of sufficient memory. At any one time all one has to store is one input line and one output line. This is accomplished as follows.

Having determined the output image height from the input height and the vertical compression factor, write a loop that repeats this number of times. Within the loop set to zero all the elements of an integer array which must be at least as long as one output line. Then, using the `fread` function described in Chapter 21, read in one input line. If the horizontal compression factor is two, then sum each pair of input values and add the sums to the output array.

The output array should be of a larger type than the input array, usually `int` rather than `char`. Obviously if the horizontal compression factor is three then sum groups of three pixels, and so on. Read and sum subsequent lines until the number of lines processed is equal to the vertical compression factor. At this point convert the sums to averages by dividing each sum by the product of the horizontal and vertical compression factors, and then write the line out. Writing the line can be done, either by calling `fputc` for each byte or by using `fwrite`. If `fwrite` is used then the averaged values will need writing to a character array before output.

Testing this program is rather difficult because it is not possible to verify that the output file is correct unless one calculates the numbers by hand. For now just be confident if the resultant image size looks about right and the contents are not all zeros. Do not forget that the averaging process may cause a few lines at the bottom of the screen to be dropped. Similarly pixels at the right edge of the image may also be omitted.

Changing image resolution

The supplied data uses 256 possible intensity values but the display can only show sixteen (assuming screen mode 10 hex is used). To get round this problem a program is required that maps the 256 different source values into 16 output values. This program will be expanded further shortly but for now it is sufficient to simply divide each input value by sixteen. Thus write a program that takes a given input file, reads the header, writes out a new header (not forgetting to change the data resolution field) and then copies each data byte after dividing by 16. Examining the first few bytes of the input and output files should show when the program is working correctly. Again do not forget to use binary mode for all the files.

Displaying the image

At last the point is reached where visible output is the target. All the tools are now available and so is the data. This program should open (in binary mode) a given file. If the open is successful then the screen should be set to mode 10 hex. Then, after waiting for a character to be entered, reset the screen mode to its original value, close the data file and exit. Test this much and when happy all is well proceed as follows.

From the image width and height determine the position of the bottom left–hand corner of the image when it is centred on the screen. Clearly if the image is too large then the program should terminate with a suitable message. If this check is done before the screen mode is changed this would be neater.

Then loop reading lines of data. Reverse each line and use the raster drawing function written earlier to write the line starting at the bottom left and plotting subsequent lines successively up the screen. The reversing is required because the satellite image is scanned from bottom right to top left, the inverse of the normal television mode. The horizontal reversal could be done equally well in the raster drawing function if required.

Grey level image

If using a VGA or compatible display then you could use screen mode 12 hex to give a 640 by 480 display (again 16 colours). Alternatively, mode 13 hex gives only 320 by 200 pixels but does allow 256 colours. Colours 16 to 31 are sixteen grey levels, and quite a realistic image can be created using these values. If an SVGA display is available then one of the higher resolution modes can be used. Unfortunately there is no standard, so consult your documentation to determine which screen mode to select. With a 256 colour, high–resolution mode, using colours 16–31 can produce impressive, almost photographic, results.

Improving the colour mapping

The program that converted the 256 different data values into 16 by simply dividing each value by 16 would be fine providing the number of each data values was approximately equal. In the real data this is not so. For the supplied data most of the data values are below 128. This results in the image created in the last section consisting of mainly eight different levels. A better mapping algorithm is required.

There are a number of different techniques for mapping data of this form; a simple binning method will be described here. That is not to say that you should not experiment with other techniques.

The binning technique requires that the source data be scanned and the number of each data value counted. These counts are then used to decide where to split the data such that each output value occurs approximately the same number of times. To explain this a bit better consider the graph shown in Fig. 37.5. This shows the cumulative counts of some imaginary input data. The cumulative count is divided into sixteen equal parts and by taking the corresponding data values the mapping from input value to output value can be found.

Fig. 37.5   Cumulative frequency graph

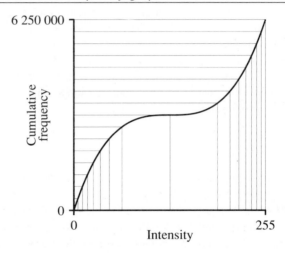

From Fig. 37.5 it can be seen that there are approximately as many data points with values in the range 0 to 10 as there are in the range 80 to 125. A straight division mapping as used earlier would waste a significant number of output colours, displaying very little data. It could be, however, that with some other data file, the only significant data are in the central region of the graph. If this were the case then alternative banding would be required.

Write a program to determine a suitable mapping, and then create a mapped datafile. The source image size and number of distinct data values can be determined from the six–byte header while the required number of output data values requires either specifying as a parameter or entering interactively. The entire data file will require reading once to determine the frequency of each data value, then the data file will require re-reading to actually do the data mapping. The standard library function `fseek` could be used to position the input file so that it can be re-read.

Having determined the required data mapping the process could be done using a series of `if` statements to check which range the current input value is, and thus the required output value. However, a much more efficient method would be to create an array equal in size to the number of different input values. In each element of this array save the value to which this element is to be mapped. Thus if data value 4 is to be mapped to 2, then the fifth array element (remember counting starts from zero) would contain the value two. This mapping technique is well worth using as the difference in speed is considerable.

Testing the program can only really be done by displaying the original result and then the one with the new mapping. Hopefully there should now be significantly more detail showing especially in the cloud regions although the colours are still not optimum.

Printing the image

If the earlier screen printing function is used to dump the image, the output is terrible, for the simple reason that the first version was really only designed for monochrome (i.e. only black or white) and could not handle shading. To enable the screen dump of an image to be achieved a new technique is required. A method of printing grey levels is required. There are many ways to do this and the one described is quite straightforward and produces reasonable results. It is known as dithering.

Given that the printer available can only either produce a dot or leave it blank, that is, is unable to generate a grey dot, then a method of simulating shading is required. Obviously by writing a pattern of dots the effect of shading can be simulated but determining what patterns to use so as to avoid obvious striping is not so easy. Another point to overcome is that if every screen pixel were printed as a cluster of sixteen dots (printed as a 4 by 4 box) which produced the required grey level, then the printed image would be considerably larger then previously, and would require viewing from a distance.

While this technique would work, the use of dithering will enable one screen pixel to be represented as one printer dot and still get the effect of grey levels. How this works is shown diagramatically in Fig. 37.6.

A hypothetical grid is placed over the data being printed. Where the data value is equal to or greater than the corresponding cell in the grid then a dot is printed at that point. Having processed each cell in the grid area the grid is moved on by the width of the grid. Having processed a complete horizontal band the grid is advanced down the image by the grid height. The grid is normally rectangular, but need not be square. The values in the grid determine the patterns generated for each data value.

Figure 37.6a shows an example grid while Fig. 37.6b – 37.6f show the patterns printed when the grid is laid over areas consisting of data values 1, 3, 5, 7, and 9. The grid is designed for an image with ten grey levels.

Fig. 37.6   Implementing dithering

<table>
<tr><td colspan="4" align="center">a</td></tr>
<tr><td>3</td><td>9</td><td>1</td><td>6</td></tr>
<tr><td>8</td><td>5</td><td>7</td><td>4</td></tr>
<tr><td>2</td><td>7</td><td>5</td><td>8</td></tr>
<tr><td>6</td><td>4</td><td>9</td><td>3</td></tr>
</table>

<table>
<tr><td colspan="4" align="center">b</td></tr>
<tr><td></td><td></td><td>*</td><td></td></tr>
<tr><td></td><td></td><td></td><td></td></tr>
<tr><td></td><td></td><td></td><td></td></tr>
<tr><td></td><td></td><td></td><td></td></tr>
</table>

<table>
<tr><td colspan="4" align="center">c</td></tr>
<tr><td>*</td><td></td><td>*</td><td></td></tr>
<tr><td></td><td></td><td></td><td></td></tr>
<tr><td>*</td><td></td><td></td><td></td></tr>
<tr><td></td><td></td><td></td><td>*</td></tr>
</table>

<table>
<tr><td colspan="4" align="center">d</td></tr>
<tr><td>*</td><td></td><td>*</td><td></td></tr>
<tr><td></td><td>*</td><td></td><td></td></tr>
<tr><td>*</td><td></td><td>*</td><td></td></tr>
<tr><td></td><td>*</td><td></td><td>*</td></tr>
</table>

<table>
<tr><td colspan="4" align="center">e</td></tr>
<tr><td>*</td><td></td><td>*</td><td>*</td></tr>
<tr><td></td><td>*</td><td>*</td><td>*</td></tr>
<tr><td>*</td><td>*</td><td>*</td><td></td></tr>
<tr><td>*</td><td>*</td><td></td><td>*</td></tr>
</table>

<table>
<tr><td colspan="4" align="center">f</td></tr>
<tr><td>*</td><td>*</td><td>*</td><td>*</td></tr>
<tr><td>*</td><td>*</td><td>*</td><td>*</td></tr>
<tr><td>*</td><td>*</td><td>*</td><td>*</td></tr>
<tr><td>*</td><td>*</td><td>*</td><td>*</td></tr>
</table>

First generate a suitable test image by using the previously written binning program to create a ten–level image. Then write a program to print the image using the dither matrix shown in Fig. 37.6a.

Implementing this algorithm is quite straightforward, primarily due to the matrix being four by four. The data can be processed by first determining which column of the dither matrix the current data byte covers and then testing each of the eight data values (one for each pin) against the corresponding dither cells. The dither matrix will require repeating to allow for the eight pixels. It takes a bit of thinking about but is remarkably simple once grasped.

Speed improvement

If happy with the speed at which the data is drawn to the screen, then this section can be omitted. This is just an extension to speed things up by about five times. If really in to bits and bytes then this is for you.

The first point to be realized is accessing the screen memory directly makes the program non-portable. On a similar computer with a different display, or display adapter, the program may well not work. This direct access should only really be done if speed is important, and the hardware is static.

Two display formats will be described. The first is that of mode 13 hex which is a standard VGA mode giving 256 colours, with a resolution of 320 pixels accross by 200 high. The second display mode will be mode 10 hex and is the one which has been used up until this point. This memory format is quite complex but the discussion will show the capabilities of C.

Directly addressing screen memory in mode 13 hex

In this mode each byte of display memory contains the colour value for one screen pixel. Colours 0–15 are the standard EGA colours, 16–31 form a 16 step grey level, and colours 32–255 are an assorted range of colour shades. The screen memory starts at address 0xA0000, and scans from the top left corner. Thus if the byte at address 0xA0000 is set to 0 then the top left pixel will be black. Setting the byte at address 0xA0001 to 15 sets the adjacent pixel to bright white. The direct addressing as described in Chapter 16 can be used, but note that no structure will be required as there is no character and attribute pair. The single byte describes the pixel completely. If you have access to a VGA display (or better) try re-writing the raster output routine to use direct addressing and compare the speeds.

If an SVGA display is available there will be other 256 colour display modes available with higher resolutions. Because there is insufficient memory space within the processor address space, it is not possible to directly address the entire display memory in one go. To allow for this, the display is split into horizontal zones and these have to be selected via the display adapter. Thus if the first zone is selected address 0xA0000 corresponds to the top left of the screen. However, if the next zone is selected, the same address now maps to a point about a quarter of the way down the screen. How these zones are defined, and how they are selected by the processor, depend on the display adapter. Details may be supplied in the adapter manual.

Directly addressing screen memory in mode 10 Hex

There is one point that requires checking before embarking on this section. In order to access the display controller at a low level it is necessary to be able to output variables to its controlling ports. Most implementations of C for the IBM PC permit this access through a library function `outp`, or some similar name. Check in the manuals for this function, and failing this, check for an index entry relating to port output. All being well you should be able to run the program shown in Fig. 37.7. This program simply sets a register to its default value but, if it can be compiled and run, then it demonstrates the required port access is available.

Fig. 37.7   Simple port access test program

```
#include <dos.h>

int main(void)
{
    outp(0x3CE, 5);
    outp(0x3CF, 0);
    return 0;
}
```

When drawing a complete horizontal line, or raster as it is called in the trade, 640 function calls are required. Each call takes the horizontal and vertical position and converts them into a memory address into which the pixel value is saved. By giving a sequence of bytes in one call the repetitive address calculation could be avoided. One calculation would be required to save the first point while the remainder would follow on sequentially. If only it were this simple. Unfortunately, screen memory and pixels are not a one–to–one mapping; indeed the mapping changes between screen modes. A complete definition of how to drive the display is given in *Programmer's Guide to the EGA and VGA Cards* by Richard F. Ferraro (1990), but be warned: it goes into over 1000 pages. The following summary applies only to screen mode 10 hex.

Screen memory format in mode 10 hex

The display memory begins at address 0xA0000. But rather than being a single block—or plane—of memory, four separate planes all begin at this address. Each plane contains one bit of each pixel value. How the card handles the overlapping blocks of memory need not concern us, just accept that it does. The top left pixel on the screen requires four bits which are taken from the left–most bit of the four bytes which all start at the same address (0xA0000). This is shown diagramatically in Fig. 37.8

Fig 37.8   Display memory format

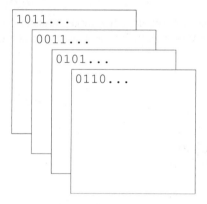

Figure 37.8 shows the four memory planes, all of which are located at address 0xA0000. The first few bits in each plane are shown and correspond to the first four pixels at the top left of the screen. The diagram shows the left–most pixel has a value of eight (1000 binary), while the second has the value three (0011 binary). In order to correctly update these memory planes one has to address the display controller at a very low level.

The mechanism of updating pixel values requires changing the internal registers of the graphics controller. These can be considered similar to the processors' registers but there are considerably more of them. To access a register requires port input/output. A port is an address that maps to a device, in our case the display controller, and is used to pass data to a device rather than memory.

In order to change a controller register the required register needs to be selected by sending its identifying number via a control port, then the value is sent, using the associated data port (usually one address higher). The function `setreg` shown in Fig. 37.9 enables a given register to be set through a specified port.

Fig. 37.9   Function to set controller register

```
setreg(int port, int reg, int value)
{
   outp(port++, reg);
   outp(port, value);
}
```

Display write modes

As if all the different screen modes were not enough, there are three different modes of writing data to the display controller. The default is write mode zero; however, for our project write mode two is considerably easier to use. To select write mode two requires the mode register to be set. This is register number five on control port 0x3CE and so can be set as follows:

```
setreg(0x3CE, 5, 2);
```

To set a specific pixel the following actions are required. Firstly the address of the byte that contains the required pixel must be calculated. As there is one bit for each of eight pixels stored in every byte in the plane memory the offset can be calculated using:

```
adr = (unsigned char _far *) 0xA0000000
         + (y * 80) + x / 8);
```

where x and y are the horizontal and vertical co-ordinates. The value of 80 comes from the total image width being 640 pixels, with eight bits per byte, hence 80 bytes per horizontal scan line.

A mask is also required that contains a single set bit corresponding to the position of the required pixel within the data byte. This can be determined as follows:

```
mask = 0x80 >> (x % 8);
```

To clarify the above, Table 37.1 lists some co-ordinates together with the corresponding memory address and mask value.

Table 37.1   Pixel co-ordinate mapping

X	Y	Address	mask (hex)
0	0	A0000	80
1	0	A0000	40
2	0	A0000	20
3	0	A0000	10
4	0	A0000	08
5	0	A0000	04
6	0	A0000	02
7	0	A0000	01
8	0	A0001	80
0	1	A0050	80

To change a pixel value the mask register needs setting to the value `mask`, the required byte requires reading, and then writing. The read is required to load the existing data into the controller, and the write is required to action the change. The value read is immaterial, it is the written value which counts. The code to perform these actions is given in Fig. 37.10 and explained in the following text.

Fig. 37.10   Writing a pixel value

```
unsigned char dummy;
unsigned char _far *adr = (unsigned char _far *)
        0xA0000000 + (y * 80) + (x / 8);
int mask = 0x80 >> (x % 8);
setreg(0x3CE, 8, mask);
dummy = *adr;
*adr = pixel;
```

The pointer variable `adr` is set to the address in the display memory that contains the required pixel. The variable `mask` is set to indicate the required bit position within the addressed byte. The mask value is put into the display controller register using the `setreg` function described earlier. Register 8 is the bit mask register that is accessed through ports 0x3CE (control) and 0x3CF (data). Assigning the variable dummy causes the controller to read all four bytes located at same display memory address into its internal registers. Storing the value causes the value to be saved in the display memory at the bit position specified by the bit mask register. If wanted, upto eight display pixels can be set at the same time by suitable values of the mask byte.

As another example, suppose the display memory were as shown in Fig. 37.8, with the pixel value being 5 and the mask register 40 hex. The result would be as shown in Fig. 37.11.

Fig. 37.11 Display memory after update

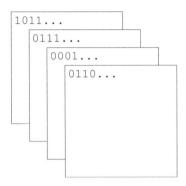

```
1011...
0111...
0001...
0110...
```

Write a function to output one horizontal line given a starting co-ordinate and an array of pixel values. This is an extension to the one written earlier using the BIOS calls though this time the code should use the register interface along the lines of that shown in Fig. 37.10. Only one memory address and byte mask calculation should be required for each function call. After that, subsequent pixels can be written by shifting the mask one bit to the right and repeating the read/write cycle. When updating the mask between pixel values, if it becomes zero then it should be reset to 0x80, and the memory address incremented.

Use the previous colour bar test program to verify the new function works. Do not forget to set the screen write mode to two after selecting the screen mode 10 hex. The new function should be significantly faster than the original version. Note, however, that it requires a compatible display and only works in screen mode 10 hex, as other modes store the pixels differently.

Conclusion

There is room for extension in all parts of this project. For example, the coastline data could be plotted using different map projections, colour added, and the major cities indicated. The image section could be enhanced by better colour mapping, and even superimposing the coastline data. The printing techniques could be improved for faster, higher–resolution printers. In short, there is a long way to go, but that is half the fun of programming.

Satellite project—setting screen mode

```
/*   Setting screen mode test program      */
/*   P. Jarvis.              13/04/1991     */

#include <stdio.h>
#include <dos.h>

int main(void);
int setmode(int);

int main(void)
{
   int mode;
   mode = setmode(16);
   getchar();
   setmode(mode);
   return 0;
}

int setmode(int mode)      /*   Set screen mode        */
{                          /*   Returns original mode  */
   int i;
   union REGS regs;
   regs.h.ah = 0x0F;
   int86(0x10, &regs, &regs);
   i = (int) regs.h.al;
   regs.h.ah = 0x00;
   regs.h.al = (unsigned char) mode;
   int86(0x10, &regs, &regs);
   return(i);
}
```

Satellite project—drawing pixels

```
/*   Drawing pixels test program   */
/*   P. Jarvis.          13/04/1991  */

#include <stdio.h>
#include <dos.h>

int main(void);
int setmode(int);
void point(int, int, int);

int main(void)
{
   int i, j, k;
   int mode;
   int line = 50;
   mode = setmode(16);        /*  Select graphics screen mode  */
   for (i=0; i<16; i++) {   /*  Loop for 16 colours          */
      for (j=0; j<10; j++) { /*  10 lines of each pixel value */
         for (k=0; k<640; k++)/*  640 pixels per line          */
            point(k, line, i);/*  Draw pixel                   */
         line++;              /* Move down one line            */
      }
   }
   getchar();                 /*  Wait for input               */
   setmode(mode);             /*  Restore original screen mode */
   return 0;
}


int setmode(int mode)       /*     Set screen mode      */
{                           /*  Returns original mode   */
   int i;
   union REGS regs;
   regs.h.ah = 0x0F;
   int86(0x10, &regs, &regs);        /*  Get current mode  */
   i = (int) regs.h.al;
   regs.h.ah = 0x00;
   regs.h.al = (unsigned char) mode;
   int86(0x10, &regs, &regs);        /*  Set new mode        */
   return(i);
}


void point(int x, int y, int pixel)  /*  Set screen pixel    */
{
   union REGS regs;
   regs.h.ah = 0x0C;
   regs.h.al = (unsigned char) pixel;  /*  Pixel value        */
   regs.h.bh = 0;                      /*  Page number        */
   regs.x.cx = x;                      /*  Horizontal position */
   regs.x.dx = y;                      /*  Vertical position   */
   int86(0x10, &regs, &regs);
}
```

Satellite project—drawing a circle

```
/*   Circle drawing test program   */
/*   P. Jarvis           14/05/1991   */

#include <stdio.h>
#include <math.h>
#include <dos.h>

int main(void);
int setmode(int);
void point(int, int, int);

int main(void)
{
   int i, j;
   int mode;
   double scale = 15.0 / 11.5;
   double angle;
   mode = setmode(16);         /*  Select graphics screen mode   */

   for (angle = 0.0; angle < 360.0; angle += 1.0) {
      i = 319 + (int) (scale * 150.0 * sin(angle * 0.0175));
      j = 174 - (int) (150.0 * cos(angle * 0.0175));
      point(i, j, 15);
   }

   getchar();                  /*  Wait for input              */
   setmode(mode);              /*  Restore original screen mode */
   return 0;
}


int setmode(int mode)    *   /*      Set screen mode    */
{                            /*  Returns original mode  */
   int i;
   union REGS regs;
   regs.h.ah = 0x0F;
   int86(0x10, &regs, &regs);         /*  Get current mode */
   i = (int) regs.h.al;
   regs.h.ah = 0x00;
   regs.h.al = (unsigned char) mode;
   int86(0x10, &regs, &regs);         /*  Set new mode      */
   return(i);
}


void point(int x, int y, int pixel)  /*  Set screen pixel     */
{
   union REGS regs;
   regs.h.ah = 0x0C;
   regs.h.al = (unsigned char) pixel;  /*  Pixel value<> */
   regs.h.bh = 0;                      /*  Page number          */
   regs.x.cx = x;                      /*  Horizontal position  */
   regs.x.dx = y;                      /*  Vertical position    */
   int86(0x10, &regs, &regs);
}
```

Satellite project—drawing lines of latitude and longitude

```c
#include <stdio.h>
#include <math.h>
#include <dos.h>

int main(void);
int setmode(int);
void point(int, int, int);


int main(void)
{
   int i, j;
   int mode;
   double scale = 15.0 / 11.5;
   double latitude, longitude;
   double sin_latitude;
   double cos_latitude;
   double sin_longitude;
   mode = setmode(16);                    /*   Select graphics screen mode   */

/*   Draw lines if longitude   */

   for (longitude=-90.0; longitude <= 90.0; longitude += 30.0) {
     sin_longitude = sin(longitude * 0.0175);
     for (latitude =- 90.0; latitude <= 90.0; latitude++) {
       i = 319 + (int) (scale * 150.0 * sin_longitude *
                    cos(latitude * 0.0175));
       j = 174 - (int) (150.0 * sin(latitude * 0.0175));
       point(i, j, 1);
     }
   }

/*   Add lines if latitude   */

   for (latitude = -60.0; latitude <= 60.0; latitude += 30.0) {
     cos_latitude = cos(latitude * 0.0175);
     sin_latitude = sin(latitude * 0.0175);
     for (longitude =- 90.0; longitude <= 90.0; longitude++) {
       i = 319 + (int) (scale * 150.0 * cos_latitude *
                    sin(longitude * 0.0175));
       j = 174 - (int) (150.0 * sin_latitude);
       point(i, j, 1);
     }
   }
   getchar();                          /*   Wait for input                */
   setmode(mode);                      /*   Restore original screen mod    */
   return 0;
}

int setmode(int mode)                  /*      Set screen mode      */
{                                      /*   Returns original mode   */
   int i;
   union REGS regs;
   regs.h.ah = 0x0F;
   int86(0x10, &regs, &regs);          /*   Get current mode   */
   i = (int) regs.h.al;
   regs.h.ah = 0x00;
   regs.h.al = (unsigned char) mode;
   int86(0x10, &regs, &regs);          /*   Set new mod      */
   return(i);
}
```

```
void point(int x, int y, int pixel)        /*  Set screen pixel  */
{
   union REGS regs;
   regs.h.ah = 0x0C;
   regs.h.al = (unsigned char) pixel;      /*  Pixel value
*/
   regs.h.bh = 0;                  /*  Page number          */
   regs.x.cx = x;                  /*  Horizontal position  */
   regs.x.dx = y;                  /*  Vertical position    */
   int86(0x10, &regs, &regs);
}
```

Satellite project—drawing a full globe

```
/*   Drawing a full globe                        */
/*   P. Jarvis.              13/04/1991           */

#include <stdio.h>
#include <math.h>
#include <dos.h>

int main(int, char **);
void exit(int);
int setmode(int);
void point(int, int, int);

int main(int argc, char *argv[])
{
   int i, j;
   int mode;
   int c[2];
   FILE *fh;
   double scale = 15.0 / 11.5;
   double latitude, longitude;
   double sin_latitude;
   double cos_latitude;
   double sin_longitude;

/*   Check and open data file   */

   if (argc != 2) {
      fprintf(stderr, "Usage:   %s   data file\n", argv[0]);
      exit(1);
   }
   fh = fopen(argv[1], "rb");
   if (fh == NULL) {
      fprintf(stderr, "Unable to open %s\n", argv[1]);
      exit(2);
   }

   mode = setmode(16);        /*   Select graphics screen mode   */

/*   Draw lines if longitude   */

   for (longitude=-90.0; longitude <= 90.0; longitude += 30.0) {
      sin_longitude = sin(longitude * 0.0175);
      for (latitude =- 90.0; latitude <= 90.0; latitude++) {
         i = 319 + (int) (scale * 150.0 * sin_longitude *
                            cos(latitude * 0.0175));
         j = 174 - (int) (150.0 * sin(latitude * 0.0175));
         point(i, j, 1);
      }
   }

/*   Add lines if latitude   */

   for (latitude = -60.0; latitude <= 60.0; latitude += 30.0) {
      cos_latitude = cos(latitude * 0.0175);
      sin_latitude = sin(latitude * 0.0175);
      for (longitude =- 90.0; longitude <= 90.0; longitude++) {
         i = 319 + (int) (scale * 150.0 * cos_latitude *
                            sin(longitude * 0.0175));
         j = 174 - (int) (150.0 * sin_latitude);
         point(i, j, 1);
      }
   }
```

```
/*   Now plot the data   */

    while (fread(c, 2, sizeof(int), fh) == 2) {
       latitude = (double) c[0] / 100.0;
       longitude = (double) c[1] / 100.0;
       if ((longitude >= -90.0) && (longitude <= 90.0)) {
          i = 319 + (int) (scale * 150.0 * cos(latitude *
                     0.0175) * sin(longitude * 0.0175));
          j = 174 - (int) (150.0 * sin(latitude * 0.0175));
          point(i, j, 15);
       }
    }

    fclose(fh);
    getchar();                      /*  Wait for input          */
    setmode(mode);                  /*  Restore original screen mode */
    return 0;
}


int setmode(int mode)        /*      Set screen mode     */
{                            /*  Returns original mode   */
    int i;
    union REGS regs;
    regs.h.ah = 0x0F;
    int86(0x10, &regs, &regs);        /*  Get current mode   */
    i = (int) regs.h.al;
    regs.h.ah = 0x00;
    regs.h.al = (unsigned char) mode;
    int86(0x10, &regs, &regs);        /*  Set new mode        */
    return(i);
}


void point(int x, int y, int pixel)   /*  Set screen pixel   */
{
    union REGS regs;
    regs.h.ah = 0x0C;
    regs.h.al = (unsigned char) pixel;   /*  Pixel value        */
    regs.h.bh = 0;                       /*  Page number        */
    regs.x.cx = x;                       /*  Horizontal position */
    regs.x.dx = y;                       /*  Vertical position   */
    int86(0x10, &regs, &regs);
}
```

Satellite project—function to write screen to Epson printer

```
/*  Print screen test program  */
/*  P. Jarvis       14/05/1993  */

#include <stdio.h>
#include <math.h>
#include <dos.h>

#define TRUE 1
#define FALSE 0

int main(void);
int setmode(int);
void point(int, int, int);
int sdump(void);

int main(void)
{
   int i, j;
   int mode;
   double scale = 15.0 / 11.5;
   double angle;
   mode = setmode(16);        /*  Select graphics screen mode  */

   for (angle = 0.0; angle < 360.0; angle += 1.0) {
      i = 319 + (int) (scale * 150.0 * sin(angle * 0.0175));
      j = 174 - (int) (150.0 * cos(angle * 0.0175));
      point(i, j, 15);
   }

   getchar();                 /*  Wait for input                */
   sdump();                   /*  Print the screen              */
   setmode(mode);             /*  Restore original screen mode  */
   return 0;
}


int setmode(int mode)        /*     Set screen mode      */
{                            /*  Returns original mode   */
   int i;
   union REGS regs;
   regs.h.ah = 0x0F;
   int86(0x10, &regs, &regs);        /*  Get current mode  */
   i = (int) regs.h.al;
   regs.h.ah = 0x00;
   regs.h.al = (unsigned char) mode;
   int86(0x10, &regs, &regs);        /*  Set new mode      */
   return(i);
}


void point(int x, int y, int pixel)  /*  Set screen pixel    */
{
   union REGS regs;
   regs.h.ah = 0x0C;
   regs.h.al = (unsigned char) pixel;  /*  Pixel value<> */
   regs.h.bh = 0;                    /*  Page number          */
   regs.x.cx = x;                    /*  Horizontal position  */
   regs.x.dx = y;                    /*  Vertical position    */
   int86(0x10, &regs, &regs);
}
```

```
int sdump(void)                    /*  Dump graphics screen to file    */
{                                  /*  Must be in screen mode 10 hex   */
   int i, j, k;
   int byte;
   FILE *fh;
   union REGS regs;

   fh = fopen("LPT1", "wb");
   if (fh == NULL)
      return FALSE;

   fprintf(fh, "\033A\010");                     /*  Set line spacing  */
   for (i = 0; i < 348; i += 8) {
      fprintf(fh, "\n\033K%c%c", 320%256, 320/256);
      for (j = 0; j < 640; j += 2) {
         byte = 0;
         for (k = i; k < i+8; k++) {
            byte <<= 1;                           /*  Make room for next bit
            regs.h.ah = 0x0D;
            regs.h.bh = 0;
            regs.x.cx = j;
            regs.x.dx = k;
            int86(0x10, &regs, &regs);            /*  Check left pixel   */
            if (regs.h.al == 0) {
               regs.h.ah = 0x0D;
               regs.h.bh = 0;
               regs.x.cx = j + 1;
               regs.x.dx = k;
               int86(0x10, &regs, &regs);         /*  Check right pixel  */
            }
            byte |= (regs.h.al) ? 1 : 0;
         }
         fprintf(byte, fh);
      }
   }
   fclose(fh);
   return TRUE;
}
```

Satellite project—raster output test program

```c
/*   Drawing rasters test program   */
/*   P. Jarvis.          13/04/1991  */

#include <stdio.h>
#include <dos.h>

int main(void);
int setmode(int);
void raster(int, int, unsigned char *, int);
void point(int, int, int);

int main(void)
{
   int i;
   int mode;
   unsigned char line[640];
   mode = setmode(16);        /*  Select graphics screen mode   */

   for (i=0; i<640; i++)      /*  Preset output array           */
      line[i] = (unsigned char) (i / 40);

   for (i=0; i<40; i++)                /*  Write raster 40 times */
      raster(0, i, line, 640);

   getchar();                 /*  Wait for input                */
   setmode(mode);             /*  Restore original screen mode  */
   return 0;
}


int setmode(int mode)       /*     Set screen mode      */
{                           /*  Returns original mode   */
   int i;
   union REGS regs;
   regs.h.ah = 0x0F;
   int86(0x10, &regs, &regs);        /*  Get current mode  */
   i = (int) regs.h.al;
   regs.h.ah = 0x00;
   regs.h.al = (unsigned char) mode;
   int86(0x10, &regs, &regs);        /*  Set new mode      */
   return(i);
}


void raster(int x, int y, unsigned char *ptr, int num)
{
   while (num-- > 0)
      point(x++, y, *(ptr++));
   return;
}


void point(int x, int y, int pixel)  /*  Set screen pixel    */
{
   union REGS regs;
   regs.h.ah = 0x0C;
   regs.h.al = (unsigned char) pixel;   /*  Pixel value       */
   regs.h.bh = 0;                       /*  Page number       */
   regs.x.cx = x;                       /*  Horizontal position */
   regs.x.dx = y;                       /*  Vertical position   */
   int86(0x10, &regs, &regs);
}
```

Satellite project—adding a header to the raw data file

```c
/*   Put header onto raw data file    */
/*   P. Jarvis                02/06/93    */

#include <stdio.h>

int main(int, char **);
int getvar(char *);
void exit(int);

int main(int argc, char *argv[])
{
    int i;
    int height;
    int width;
    int resolution;
    long count = 0L;
    FILE *fin, *fout;
    unsigned char line[1024];

    if (argc != 3) {
        fprintf(stderr, "Usage:   %s input_file output_file\n",
                argv[0]);
        exit(1);
    }

    fin = fopen(argv[1], "rb");
    if (fin == NULL) {
        fprintf(stderr, "Unable to open %s for input\n", argv[1]);
        exit(2);
    }

    fout = fopen(argv[2], "wb");
    if (fout == NULL) {
        fprintf(stderr, "Unable to open %s for output\n",
                argv[2]);
        exit(3);
    }

/*   Ask for the image details   */

    width = getvar("image width");
    height = getvar("image height");
    resolution = getvar("image resolution");

/*   Write the image header   */

    fputc(width/256, fout);
    fputc(width%256, fout);
    fputc(height/256, fout);
    fputc(height%256, fout);
    fputc(resolution/256, fout);
    fputc(resolution%256, fout);

/*   Now copy the data   */

    while ((i = fread(line, 1, 1024, fin)) > 0) {
        if (fwrite(line, 1, i, fout) != i) {
            fprintf(stderr, "Data copy failed to complete\n");
            exit(4);
        }
        count += (long) i;
    }
```

```
/*   Check length matches dimensions given   */

   if (count != (long) width * (long) height) {
      fprintf(stderr, "Data size does not match dimensions\n");
      exit(5);
   }

   fclose(fin);
   fclose(fout);
   return 0;
}

int getvar(char *prompt)
{
   int result;
   char inp[80];
   do {
      printf("Please enter the %s: ", prompt);
      if (fgets(inp, 80, stdin) != NULL) {
         if ((sscanf(inp, "%d", &result) == 1) && (result > 0))
            break;
      }
   } while (1);
   return result;
}
```

Satellite project—scaling the image

```c
/*   Scale given image by integral amount   */
/*   P. Jarvis                      02/06/93   */

#include <stdio.h>
#include <stdlib.h>

int main(int, char **);
int getvar(char *);
void exit(int);

int main(int argc, char *argv[])
{
   int i, j, k;
   int height_old;              /*   Original image height       */
   int width_old;               /*   Original image width        */
   int height_new;              /*   New image height            */
   int width_new;               /*   New image width             */
   int resolution;              /*   Image resolution            */
   int scale;                   /*   Number of cells to average  */
   int vertical_scale;          /*   Vertical scale factor       */
   int horizontal_scale;        /*   Horizontal scale factor     */
   FILE *fin, *fout;
   unsigned char line[8];
   unsigned char *inline;
   unsigned int *outline;

   if (argc != 3) {
      fprintf(stderr, "Usage:   %s input_file output_file\n",
              argv[0]);
      exit(1);
   }

   fin = fopen(argv[1], "rb");
   if (fin == NULL) {
      fprintf(stderr, "Unable to open %s for input\n", argv[1]);
      exit(2);
   }

   fout = fopen(argv[2], "wb");
   if (fout == NULL) {
      fprintf(stderr, "Unable to open %s for output\n",
              argv[2]);
      exit(3);
   }

/*   Read source image details   */

   if (fread(line, 1, 6, fin) != 6) {
      fprintf(stderr, "Unable to load source header\n");
      exit(4);
   }
   width_old = line[0] * 256 + line[1];
   height_old = line[2] * 256 + line[3];
   resolution = line[4] * 256 + line[5];
   printf("\nSource image details:\n\n");
   printf("Width       %d\nHeight       %d\nResolution %d\n\n",
              width_old, height_old, resolution);

/*   Ask for the compression factors   */

   horizontal_scale = getvar("horizontal compression factor");
   vertical_scale = getvar("vertical compression factor");
   scale = horizontal_scale * vertical_scale;
/*   Scale and display new image size   */
```

```
    width_new = width_old / horizontal_scale;
    height_new = height_old / vertical_scale;
    printf("\nDestination image details:\n\n");
    printf("Width        %d\nHeight        %d\nResolution %d\n\n",
                          width_new, height_new, resolution);

/*  Allocate space for input and output data   */

    inline = malloc(width_old * sizeof(unsigned char));
    outline = malloc(width_new * sizeof(unsigned int));
    if ((inline == NULL) || (outline == NULL)) {
      fprintf(stderr, "Unable to allocate space for copy\n");
      exit(5);
    }

/*  Write the image header   */

    fputc(width_new/256, fout);
    fputc(width_new%256, fout);
    fputc(height_new/256, fout);
    fputc(height_new%256, fout);
    fputc(resolution/256, fout);
    fputc(resolution%256, fout);

/*  Now copy the data   */

    for (i=0; i < height_new; i++) {
      for (j=0; j < width_new; j++)
        outline[j] = 0;
      for (j=0; j < vertical_scale; j++) {
        if (fread(inline, 1, width_old, fin) != width_old) {
          fprintf(stderr, "Read error on source data\n");
          exit(6);
        }
        for (k=0; k < width_old; k++)
          outline[k/horizontal_scale] += (unsigned int) inline[k];
      }
      for (j=0; j < width_new; j++)
        inline[j] = (unsigned char) (outline[j] / scale);
      if (fwrite(inline, 1, width_new, fout) != width_new) {
        fprintf(stderr, "Unable to complete data copy\n");
        exit(7);
      }
    }
    fclose(fin);
    fclose(fout);
    return 0;
}

int getvar(char *prompt)
{
    int result;
    char inp[80];
    do {
      printf("Please enter the %s: ", prompt);
      if (fgets(inp, 80, stdin) != NULL) {
        if ((sscanf(inp, "%d", &result) == 1) && (result > 0))
          break;
      }
    } while (1);
    return result;
}
```

Satellite project—changing the image resolution (first try)

```
/*   Changing the image resolution   */
/*   P. Jarvis                02/06/93   */

/*   Simple conversion from 256 levels    */
/*   to 16 levels by dividing each value   */
/*   by 16.                                */

#include <stdio.h>
#include <stdlib.h>

int main(int, char **);
void exit(int);

int main(int argc, char *argv[])
{
   int i;
   int height;                /*   Image height      */
   int width;          /*   Image width      */
   int resolution;     /*   Image resolution    */
   FILE *fin, *fout;
   unsigned char line[6];

   if (argc != 3) {
      fprintf(stderr, "Usage:   %s input_file output_file\n",argv[0]);
      exit(1);
   }

   fin = fopen(argv[1], "rb");
   if (fin == NULL) {
      fprintf(stderr, "Unable to open %s for input\n", argv[1]);
      exit(2);
   }

   fout = fopen(argv[2], "wb");
   if (fout == NULL) {
      fprintf(stderr, "Unable to open %s for output\n", argv[2]);
      exit(3);
   }

/*   Read source image details   */

   if (fread(line, 1, 6, fin) != 6) {
      fprintf(stderr, "Unable to load source header\n");
      exit(4);
   }
   width = line[0] * 256 + line[1];
   height = line[2] * 256 + line[3];
   resolution = line[4] * 256 + line[5];

   printf("\nSource image details:\n\n");
   printf("Width      %d\nHeight      %d\nResolution %d\n\n",
            width, height, resolution);
   resolution /= 16;

/*   Write the image header   */

   fputc(width/256, fout);
   fputc(width%256, fout);
   fputc(height/256, fout);
   fputc(height%256, fout);
   fputc(resolution/256, fout);
   fputc(resolution%256, fout);
```

```
/*   Now  copy  the  data   */

   while  ((i=fgetc(fin))  !=  EOF)  {
      i  /=  16;
      if  (fputc(i,  fout)  !=  i)  {
         perror("Copy  failed");
         exit(5);
      }
   }

   fclose(fin);
   fclose(fout);
   return  0;
}
```

Satallite project—displaying a 16 level image

```
/*   Displaying a 16 level image   */
/*   P. Jarvis               02/06/93   */

#include <stdio.h>
#include <stdlib.h>
#include <dos.h>

int main(int, char **);
void exit(int);
int setmode(int);
void raster(int, int, unsigned char *, int);
void point(int, int, unsigned char);

int main(int argc, char *argv[])
{
   int i;
   int mode;           /*   Screen mode          */
   int x, y;           /*   Image position       */
   int height;             /*   Image height       */
   int width;          /*   Image width        */
   int resolution;     /*   Image resolution     */
   FILE *fh;
   unsigned char line[640];

   if (argc != 2) {
      fprintf(stderr, "Usage:   %s input_file\n", argv[0]);
      exit(1);
   }

   fh = fopen(argv[1], "rb");
   if (fh == NULL) {
      fprintf(stderr, "Unable to open %s for input\n", argv[1]);
      exit(2);
   }

/*   Read source image details   */

   if (fread(line, 1, 6, fh) != 6) {
      fprintf(stderr, "Unable to load source header\n");
      exit(3);
   }
   width = line[0] * 256 + line[1];
   height = line[2] * 256 + line[3];
   resolution = line[4] * 256 + line[5];

   if (resolution > 16) {
      fprintf(stderr, "Image resolution is larger than 16\n");
      exit(4);
   }

/*   calculate the image position   */

   if ((width > 640) || (height > 350)) {
      fprintf(stderr, "The image is larger than the screen
size\n");
      exit(5);
   }
   x = 640 - (640 - width) / 2;
   y = 350 - (350 - height) / 2;
```

```
/*   Now display the data   */

   mode = setmode(16);
   for (i=0; i<height; i++) {
     if (fread(line, 1, width, fh) != width)
       break;
     raster(x, y--, line, width);
   }
   getchar();
   setmode(mode);

   fclose(fh);
   return 0;
}


int setmode(int mode)        /*  Set screen mode        */
{                            /*  Returns original mode  */
   int i;
   union REGS regs;
   regs.h.ah = 0x0F;
   int86(0x10, &regs, &regs);        /*  Get current mode */
   i = (int) regs.h.al;
   regs.h.ah = 0x00;
   regs.h.al = (unsigned char) mode;
   int86(0x10, &regs, &regs);        /*  Set new mode     */
   return(i);
}


void raster(int x, int y, unsigned char *ptr, int num)
{
   while (num-- > 0)
     point(x--, y, *(ptr++));
   return;
}


void point(int x, int y, unsigned char pixel)   /*  Set screen
pixel */
{
   union REGS regs;
   regs.h.ah = 0x0C;
   regs.h.al = pixel;            /*  Pixel value         */
   regs.h.bh = 0;                /*  Page number         */
   regs.x.cx = x;                /*  Horizontal position */
   regs.x.dx = y;                /*  Vertical position   */
   int86(0x10, &regs, &regs);
}
```

Satellite project—changing the image resolution (second try)

```
/*   Change image resolution using binning */
/*   P. Jarvis                    02/06/93 */

#include <stdio.h>

int main(int, char **);
int getvar(char *);
void exit(int);

int main(int argc, char *argv[])
{
   int i, j;
   long step;
   int height;
   int width;
   int old_resolution;
   int new_resolution;
   FILE *fin, *fout;
   static long counts[256];
   unsigned char map[256];

   if (argc != 3) {
      fprintf(stderr, "Usage:   %s input_file output_file\n",
             argv[0]);
      exit(1);
   }

   fin = fopen(argv[1], "rb");
   if (fin == NULL) {
      fprintf(stderr, "Unable to open %s for input\n", argv[1]);
      exit(2);
   }

   fout = fopen(argv[2], "wb");
   if (fout == NULL) {
      fprintf(stderr, "Unable to open %s for output\n",
             argv[2]);
      exit(3);
   }

/*   Get the old image details   */

   if (fread(map, 1, 6, fin) != 6) {
      fprintf(stderr, "Unable to load the image header\n");
      exit(4);
   }
   width = map[0]  * 256 + map[1];
   height = map[2] * 256 + map[3];
   old_resolution = map[4]  * 256 + map[5];

/*   Ask for the new image resolution   */

   new_resolution = getvar("Required number of levels");

/*   Write the image header   */

   fputc(width/256,  fout);
   fputc(width%256,  fout);
   fputc(height/256,  fout);
   fputc(height%256,  fout);
   fputc(new_resolution/256,  fout);
   fputc(new_resolution%256,  fout);
```

```
/*   Count frequency of each input data value   */

   for (i=0; i<old_resolution; i++)
      counts[i] = 0L;

   while ((i=fgetc(fin)) != EOF)
      counts[i]++;

/*   Build cumulative frequency table   */

   for (i=1; i<old_resolution; i++)
      counts[i] += counts[i-1];

/*   Build mapping table   */

   j = 0;
   for (i=0; i<new_resolution; i++) {
      step = ((long) (i + 1) * (long) width * (long) height) /
(long) new_resolution;
      while ((counts[j] <= step) && (j < old_resolution)) {
         map[j] = (unsigned char) i;
         j++;
      }
   }

/*   Print out the map just for interest   */

   for (i=0; i<old_resolution; i++)
      printf("%5d", map[i]);

/*   Now map the data   */

   fseek(fin, 6L, SEEK_SET);              /*   Rewind to start of data   */
   while ((i=fgetc(fin)) != EOF) {
      i = map[i];
      if (fputc(i, fout) != i) {
         fprintf(stderr, "Failure to complete new data file\n");
         exit(5);
      }
   }

   fclose(fin);
   fclose(fout);
   return 0;
}


int getvar(char *prompt)
{
   int result;
   char inp[80];
   do {
      printf("Please enter the %s: ", prompt);
      if (fgets(inp, 80, stdin) != NULL) {
         if ((sscanf(inp, "%d", &result) == 1) && (result > 0))
            break;
      }
   } while (1);
   return result;
}
```

Satellite project—printing an image using dithering

```
/*   Image output to Epson printer using dithering     */
/*   P. Jarvis                          02/06/1993     */

#include <stdio.h>

#define  TRUE  1
#define  FALSE  0

void main(void);
void exit(int);
void pline(unsigned char *, int);

/*   Test program to generate a ten level grey scale   */
/*   and use the dither print function to output it     */
/*   to an Epson printer.                               /*

void main(void)
{
  int i;
  unsigned char line[640];
  for (i=0;  i<640;  i++)
    line[i] = (unsigned char) (i / 64);
  for (i=0;  i<350;  i++)
    pline(line,  640);
  pline(line,  -1);
  return 0;
}



/*   Function to write one raster to the file "image.prn"   */
/*   using a ten level four by four dither matrix. On the    */
/*   first call the output file is automatically opened.     */
/*   Calling this function with a negative data count causes */
/*   the output file to be closed.                           */

void pline(unsigned char *data, int num)
{
  int i;
  static FILE *fh;
  static unsigned char map[16] = {   3,  9,  1,  6,  8,  5,  7,  4,
                                     2,  7,  5,  8,  6,  4,  9,  3};
  unsigned char *dither;
  static int first = TRUE;
  static int ind;
  static int width;
  static unsigned char line[816];

/*   Close file is special request   */

  if (num < 0) {
    fclose(fh);
    first = TRUE;
    return;
  }
```

```
/*   Open file if first call    */

   if (first) {
      fh = fopen("image.prn", "wb");
      if (fh == NULL) {
        fprintf(stderr, "Unable to open image.prn for output\n");
        exit(1);
      }
      width = (num > 816) ? 816 : num;
      ind = 0;
      fprintf(fh, "\033@\033A\010");
      first = FALSE;
   }

/*   clear destination buffer if start of byte    */

   if (ind == 0) {
      for (i=0; i<width; i++)
        line[i] = 0;
   }

/*   determine dither pattern address for line    */

   dither = map + (ind % 4) * 4;

/*   Add this set of data    */

   if (num > width)
      num = width;

   for (i=0; i<num; i++) {
      line[i] <<= 1;
      if (data[i] >= dither[i%4])
        line[i] = line[i] | 1;
   }

/*   If last bit in byte write out the line    */

   if (++ind == 8) {
      fprintf(fh, "\n\033K%c%c", width%256, width/256);
      for (i=0; i<width; i++)
        fputc(line[i], fh);
      ind = 0;
   }
}
```

Satellite project—drawing pixels faster (mode 13 hex)

```
/*   Test program to write directly to screen memory    */
/*   using VGA graphics mode 13 hex.                     */
/*   P. Jarvis                          6/6/1993         */

#include <stdio.h>
#include <dos.h>

int main(void);
int setmode(int);
void raster(int, int, unsigned char *, int);

int main(void)
{
   int i;
   int mode;
   unsigned char line[320];

/*   Select mode 13 hex (320 x 200)   */

   mode = setmode(0x13);

/*   Preset pixel value array to give 16 colour bars   */

   for (i=0; i<320; i++)
      line[i] = (unsigned char) (i / 20);

/*   Fill all 200 scan lines with colour bars   */

   for (i=0; i<200; i++)
      raster(0, i, line, 640);

/*   Wait for input then return to normal mode   */

   getchar();
   setmode(mode);
   return 0;
}

int setmode(int mode)
{
   int i;
   union REGS regs;
   regs.h.ah = 0x0F;
   int86(0x10, &regs, &regs);             /*   Get current mode   */
   i = (int) regs.h.al;
   regs.h.ah = 0x00;
   regs.h.al = (unsigned char) mode;
   int86(0x10, &regs, &regs);             /*   Set new mode   */
   return i;
}


void raster(int x, int y, unsigned char *data, int num)
{
   unsigned char _far *ptr;
   ptr = (unsigned char _far *) 0xA0000000 + y * 320 + x;
   while (num-- > 0)
      *(ptr++) = *(data++);
   return;
}
```

Satellite project—drawing pixels faster (mode 10 hex)

```
/*   Faster screen update program   */
/*   P. Jarvis.          17/04/1991  */

#include <stdio.h>
#include <dos.h>
#include <conio.h>

int main(void);
void setreg(int, int, int);
void raster(int, int, unsigned char *, int);
int setmode(int);

int main(void)
{
   int i, mode;
   unsigned char line[640];

/*   Select mode 16 (640 x 350)   */

   mode = setmode(16);

/*   Select write mode 2   */

   setreg(0x3CE, 5, 2);      /*   Enable set/reset register   */

/*   Preset pixel value array to give 16 colour bars   */

   for (i=0; i<640; i++)
      line[i] = (unsigned char) (i / 40);

/*   Fill all 350 scan lines with colour bars   */

   for (i=0; i<350; i++)
      raster(0, i, line, 640);

/*   Wait for input then return to normal mode   */

   getchar();
   setmode(mode);
   return 0;
}




int setmode(int mode)      /*      Set screen mode      */
{                          /*   Returns original mode   */
   int i;
   union REGS regs;
   regs.h.ah = 0x0F;
   int86(0x10, &regs, &regs);      /*   Get current mode   */
   i = (int) regs.h.al;
   regs.h.ah = 0x00;
   regs.h.al = (unsigned char) mode;
   int86(0x10, &regs, &regs);      /*   Set new mode      */
   return(i);
}
```

```
void raster(int x, int y, unsigned char *line, int num)
{
   int i;
   unsigned char dummy;
   unsigned char far *ptr = (unsigned char far *)
                  (0xA0000000 + (y * 80) + (x / 8));
   int mask = 0x80 >> (x % 8);
   for (i=0; i<num; i++) {
      setreg(0x3CE, 8, mask);
      dummy = *ptr;
      *ptr = (unsigned char) line[i];
      mask >>= 1;
      if (mask == 0) {
         mask = 0x80;
         ptr++;
      }
   }
}


void setreg(int port, int reg, int value)
{
   outp(port, reg);
   outp(port+1, value);
}
```

Appendix A

IBM/PC character set

	0	16	32	48	64	80	96	112	128	144	160	176	192	208	224	240
0		►		0	@	P	`	p	Ç	É	á	░	└	╨	α	≡
1	☺	◄	!	1	A	Q	a	q	ü	æ	í	▒	┴	╤	ß	±
2	☻	↕	"	2	B	R	b	r	é	Æ	ó	▓	┬	╥	Γ	≥
3	♥	‼	#	3	C	S	c	s	â	ô	ú	│	├	╙	π	≤
4	♦	¶	$	4	D	T	d	t	ä	ö	ñ	┤	─	╘	Σ	⌠
5	♣	§	%	5	E	U	e	u	à	ò	Ñ	╡	┼	╒	σ	⌡
6	♠	▬	&	6	F	V	f	v	å	û	ª	╢	╞	╓	µ	÷
7	•	↨	'	7	G	W	g	w	ç	ù	º	╖	╟	╫	τ	≈
8	▪	↑	(	8	H	X	h	x	ê	ÿ	¿	╕	╚	╪	Φ	°
9	○	↓	)	9	I	Y	i	y	ë	Ö	⌐	╣	╔	┘	Θ	∙
10	◙	→	*	:	J	Z	j	z	è	Ü	¬	║	╩	┌	Ω	·
11	♂	←	+	;	K	[	k	{	ï	¢	½	╗	╦	█	δ	√
12	♀	∟	,	<	L	\	l	|	î	£	¼	╝	╠	▄	∞	^{n}
13	♪	↔	−	=	M	]	m	}	ì	¥	¡	╜	═	▌	ø	²
14	♫	▲	.	>	N	^	n	~	Ä	₧	«	╛	╬	▐	ε	■
15	☼	▼	/	?	O	_	o	⌂	Å	ƒ	»	┐	╧	▀	∩	

Shaded charaters are extensions to the ASCII definition.

Appendix B

Operator precedence

Operator(s)	Associativity
( ) [ ] -> .	Left to Right
! ~ ++ -- - (type) sizeof	Right to left †
* / %	Left to right
+ -	Left to right
<< >>	Left to right
< <= > =>	Left to right
== !=	Left to right
&	Left to right
^	Left to right
|	Left to right
&&	Left to right
||	Left to right
?:	Right to left
= += -= /= *= %= &= |= ^=	Right to left
,	Left to right

† When * and & are used as pointer references they are processed at this level.

Appendix C

IBM/PC screen modes

Mode	Resolution	Colours	Address	Hardware
00h	40 x 25 text	Mono	B800	
01h	40 X 25 text	16	B800	
02h	80 x 25 text	Mono	B800	
03h	80 x 25 text	16	B800	
04h	320 x 200 graph	4	B800	CGA
05h	320 x 200 graph	4 (grey)	B800	CGA
06h	640 x 200 graph	Mono	B800	CGA
07h	80 x 25 text	Mono	B000	MDA†
08h	20 x 25 text	16	B800	PC junior
09h	40 x 25 text	16	B800	PC junior
0Ah	80 x 25 text	4	B800	PC junior
0Bh	Reserved			
0Ch	Reserved			
0Dh	320 x 200 graph	16	A000	EGA/VGA
0Eh	640 x 200 graph	16	A000	EGA/VGA
0Fh	640 x 350 graph	Mono	A000	EGA/VGA
10h	640 x 350 graph	16	A000	EGA/VGA
11h	640 x 480 graph	2	A000	VGA
12h	640 x 480 graph	16	A000	VGA
13h	320 x 200 graph	256	A000	VGA

† Default for Hercules cards

Appendix D

Function prototypes

This Appendix lists the function prototypes of all the standard library functions mentioned in the text. They are grouped by functionality, that is, all string functions are together, as are all the maths functions. Two terms used within the prototypes need explaining. The symbol . . . indicates a variable number of arguments are required, no arguments also being valid. With these functions there is some other indication of how many arguments are to be processed. For example, the format string in the printf function indicates the number of variables to be printed. The second term introduced here is the keyword const. This has not been discussed in the text, and can be taken here as meaning that the function does not change this argument. Note that this is not a sufficient definition of the const keyword, which is beyond the scope of this text.

Character input and output

```
int     fgetc(FILE *);
char *  fgets(char *, int, FILE*);
int     fputs(const char *, FILE *);
int     fputc(int, FILE *);
int     getchar(void);
char *  gets(char *);
int     putchar(int);
int     puts(const char *);
```

Formatted input and output

```
int     fprintf(FILE *, const char *, ...);
int     fscanf(FILE *, const char *, ...);
int     printf(const char *, ...);
int     scanf(const char *, ...);
int     sprintf(char *, const char *, ...);
int     sscanf(const char *, const char *, ...);
```

File handling

```
int     fclose(FILE *);
FILE *  fopen(const char *, const char *);
int     fseek(FILE *, long, int);
long    ftell(FILE *);
```

Block input and output

```
size_t fread(void *, size_t, size_t, FILE *);
size_t fwrite(const void *, size_t, size_t,
               FILE *);
```

Memory allocation

```
void    free(void *);
void *  malloc(size_t);
```

Mathematical

```
double atan2(double, double);
double cos(double);
double fabs(double);
double log(double);
double pow(double);
double sin(double);
double tan(double);
```

String handling

```
char * strcat(char *, const char *);
char * strchr(const char *, int);
int    strcmp(const char *, const char *);
char * strcpy(char *, const char *);
int    strlen(const char *);
```

Character testing

```
int     isalpha(int);
int     isdigit(int);
int     islower(int);
int     ispunct(int);
int     isspace(int);
int     isupper(int);
```

Character conversion

```
int     tolower(int);
int     toupper(int);
```

String conversion

```
double  atof(const char *);
int     atoi(const char *);
long    atol(const char *);
```

Error handling

```
int     feof(FILE *);
int     ferror(FILE *);
void    perror(const char *);
```

Miscellaneous

```
int     abs(int);
void    exit(int);
long    labs(long);
```

Common DOS extensions

```
int     getch(void);
int     kbhit(void);
int     intdos(union REGS *, union REGS *);
int     int86(int, union REGS *, union REGS *);
```

Appendix E

ANSI escape sequences

This section lists some of the character sequences which are processed by the ANSI.SYS device driver. They are used to control how and where text is written to the screen. In general these sequences follow the ANSI X3-64 standard. The ANSI abreviation for each sequence is listed in the title lines. Note that there must not be any spaces in the sequence, and that ESC means the 'escape code' (27 decimal).

ED   Erase display

```
ESC [ 2 J
```

The erase display sequence clears the screen and positions the cursor at the home position. This is the top left corner of the screen (line 1, column 1).

EL   Erase line

```
ESC [ 2 K
```

The erase line sequence erases (i.e. sets to blank) all the character positions on the current line, from the cursor position to the right edge. Characters to the left of the cursor are not changed.

CUP   Cursor position

```
ESC [ n₁ ; n_c H
```

The cursor position sequence moves the cursor to the given line and column position. The decimal number n_1 is the line number, and n_c is the column number. Note that both numbers start at one. The ANSI standard also defines a set horizontal and vertical position sequence which has the same effect. This sequence (known as HVP) has the following format:

```
ESC [ n₁ ; n_c f
```

If the line and column numbers are omitted then values of one are assumed. The following will position the cursor on the top left corner of the screen:

```
ESC [ H
```

CUU   Cursor up

```
ESC  [  n_x  A
```

This sequence moves the cursor up by n_x lines. The cursor stays in the same column and stops when reaching the top line. The screen does not scroll. If n_x is omitted then a value of one is assumed.

CUD   Cursor down

```
ESC  [  n_x  B
```

This sequence moves the cursor down by n_x lines. The cursor stays in the same column, and stops on reaching the bottom line. The screen does not scroll. If n_x is omitted a value of one is assmed.

CUF   Cursor forward

```
ESC  [  n_x  C
```

This sequence moves the cursor to the right by n_x columns. The cursor stays on the same line and stops when reaching the right margin. If n_x is omitted then a value of one is assumed.

CUB   Cursor backward

```
ESC  [  n_x  D
```

This sequence moves the cursor to the left by n_x columns. The cursor stays on the same line, and stops on reaching the left margin.. If n_x is omitted a value of one is assumed. Note that characters on the line are not erased; this function simply moves the cursor.

SCP   Save cursor position

```
ESC  [  s
```

This sequence saves the current cursor position. It is used in conjuntion with the following sequence which restores the cursor position.

RCP   Restore cursor position

```
ESC  [  u
```

This sequence returns the cursor to the position saved by the 'save cursor position' function. If a position has not been saved then a restore is not defined.

SGR   Set graphics rendition

```
ESC [ Ps ; ... ; Ps m
```

This sequence is used to define how text is to be displayed. For example the colour and style of the text can be set. The sequence can take an arbitrary number of parameters, each one adding to the previous selection. A parameter of zero resets everything to a default normal value. The other permitted values are listed in the following table.

Parameter	Action
1	Bold
3	Italic
5	Blink
7	Reverse video
30	Black foreground
31	Red foreground
32	Green foreground
33	Yellow foreground
34	Blue foreground
35	Magenta foregroud
36	Cyan foreground
37	White foreground
40	Black background
41	Red background
42	Green background
43	Yellow background
44	Blue background
45	Magenta backgroud
46	Cyan background
47	White background

As an example the following sequence sets the display so that all subsequent text is written in bold characters which are yellow on a blue background.

```
ESC [ 0 ; 1 ; 33 ; 44 ; m
```

Appendix F

TIFF tag values

Tag	Type	Meaning
255	3	Subfield designator
256	3	Image width
257	3	Image height
258	3	Number of bits per pixel
259	3	Compression
262	3	Photometric interpretation
263	3	Thresholding
266	3	Fill order
273	4	Strip offsets
274	3	Orientation
277	3	Number of samples per pixel
278	4	Number of rows per strip
279	4	Number of bytes per strip
280	3	Minimum pixel value
281	3	Maximum pixel value
282	5	Horizontal resolution
283	5	Vertical resolution
284	3	Planar configuration

Appendix G

Screen BIOS functions

The screen is addressed through BIOS interrupt 10 hex (16 decimal). The following are some of the major features provided by this interrupt.

Set video mode

```
AH = 0
AL = Required mode (See Appendix C)
```

Set cursor attributes

```
AH = 1
CH = Cursor start line (bits 0-4)
     Bits 5 & 6 attributes:
     00 = normal
     01 = invisible
     10 = slow blink
     11 = fast blink
CL = Cursor end line (bits 0-4)
```

Set cursor position

```
AH = 2
BH = Display page
DH = Row number
DL = Column number
```

Read cursor position

```
AX = 3
BH = Display page
```

Returns:

```
DH = Row number
DL = Column number
```

Read light pen position

```
AH =  4
```

Returns:

```
AH =  0 (light pen not active)
AH =  1 (light pen active)
DH =  Row number
DL =  Column number
```

Select display page

```
AH =  5
AL =  Required page
```

Scroll up

```
AH =  6
AL =  Number of lines to scroll
BH =  attribute for blanked lines
CH =  Row number of upper left corner
CL =  Column number of upper left corner
DH =  Row number of lower right corner
DL =  Column number of lower right corner
```

Scroll down

```
AH =  7
```
Other registers as for scroll up

Read character at cursor position

```
AH =  8
BH =  Display page
```

Returns:

```
AL =  Character
AH =  Attributes
```

Write character and attribute at cursor position

```
AH  =   9
AL  =   Character
BH  =   Display page
BL  =   Attributes
CX  =   Number of times to write character
```

Write character at cursor position

```
AH  =   0A
AL  =   Character
BH  =   Display page
CX  =   Number of times to write character
```

Set colour palette

```
AH  =   0B
BH  =   0
        BL  =   Border colour
BH  =   1
        BL  =   Palette number
```

Write screen pixel

```
AH  =   0C
AL  =   Colour
BH  =   Display page
CX  =   Column
DX  =   Row
```

Read screen pixel

```
AH  =   0D
BH  =   Display page
CX  =   Column
DX  =   Row
```

Returns:

```
AL  =   Pixel value
```

Write character and advance cursor

```
AH  =   0E
AL  =   Character
BH  =   Display page
BL  =   Foreground colour
```

Get current screen mode

```
AH  =   0F
```

Returns:

```
AL  =   Screen mode (see Appendix C)
BL  =   Current display page
```

Set palette register

```
AH  =   10H
AL  =   0
BL  =   Palette register
BH  =   Colour value
```

Notes

The above is by no means a complete list of all the functions available, neither are all the possible options given for the listed functions. For full details consult a reference manual, such as *Programmer's guide to the EGA and VGA Cards* by Richard F. Ferraro (1990), or the Interrupt list by Ralf Brown which is available from bulletin boards and networks.

Appendix H

Mouse BIOS interface

The mouse interface is via BIOS Interrupt 33 hex (51 decimal). As before, this is not a complete list of the features available, but does summarize the more useful ones.

Reset and check status

```
AX = 0
```

Returns:

```
AX =     0  No mouse available
        -1  Mouse ready
BX =    -1  Two button mouse
         0  Other than two buttons
```

Show mouse cursor

```
AX = 1
```

Hide mouse cursor

```
AX = 2
```

Get mouse cursor position

```
AX = 3
```

Returns:

```
BX =  Button status
      (bit 0 = left button, bit 1 = right)
CX =  Column position
DX =  Row position
```

Set mouse cursor position

$$
\begin{aligned}
AX &= 4 \\
CX &= \text{Column position} \\
DX &= \text{Row position}
\end{aligned}
$$

Notes

The mouse BIOS routines are accessed using the interface routine `int86`. As an example the following code initializes the mouse driver:

```
union REGS regs;
regs.x.ax = 0;
int86(0x33, &regs, &regs);
if (regs.x.ax == 0)
   printf("Mouse driver not installed\n");
else
   printf("Mouse ready\n");
```

After initialization the mouse cursor is not visible until the show cursor function is called. The hide cursor function removes the cursor from the display. If the show cursor function is called more than once then a similar number of calls to the hide function are required before the cursor is removed.

Appendix I

Exercise answers

Chapter 2

1 Comments:

```
/*   Program to print 'Hello World'   */
/*   P. Jarvis.             24/11/1990   */
```

Function definition:

```
main()
{
   printf("Hello World\n");
}
```

Pre-processor directive:

```
#include <stdio.h>
```

Statement:

```
printf("Hello World\n");
```

Chapter 3

1 char
int
float
double

2 17 octal is 15 decimal
17 decimal is obvious
17 hex is 23 decimal
100 octal is 64 decimal
100 hex is 256 decimal
321 octal is 209 decimal
321 hex is 801 decimal

3   `unsigned int` would be 0 – 65 535 (using two byte integers).

4   0x3FF80000 is the internal representation of 1.9375
The internal form of 6.0 is 0x40C00000

Chapter 4

1
```
3.0     double
2L      long int
'A'     char (or int)
3.5L    long double
0x11    int (or char)
4.1F    float
```

2
```
11L     eleven
015     thirteen (leading zero causes octal format)
0x10    sixteen (hexadecimal format)
```

Chapter 5

1   `2pi` is illegal as it starts with a number
`bill.` contains an invalid character
`register` is illegal as it is a reserved word

2   `i` and `k` are both preset to zero as both are static
the value of `j` is undefined

3   Yes, static variables maintain their values between function calls.

4   Yes, global variables are also static.

Chapter 6

1
```
3
17.5
15
3
0
10
2                (Machine dependent, but generally 2 for an IBM/PC)
```

Chapter 7

1
```
/*   Program to print the size of variables   */

#include <stdio.h>

main()
{
  printf("Size of int is %d bytes\n",
             (int) sizeof(int));
  printf("Size of short is %d bytes\n",
             (int) sizeof(short int));
  printf("Size of long is %d bytes\n",
             (int) sizeof(long));
  printf("Size of float is %d bytes\n",
             (int) sizeof(float));
  printf("Size of double is %d bytes\n",
             (int) sizeof(double));
  printf("Size of long double is %d bytes\n",
             (int) sizeof(long double));
  printf("Size of char is %d bytes\n",
             (int) sizeof(char));
}
```

Chapter 9

1
```
#include <stdio.h>
main()
{
  int i = 3;
  if (i < 0)
    printf("i is negative\n");
  else
    printf("i is positive\n");
}
```

2
```
if (j < k)
  i = j;
else
  i = k;
```

3
```
i = (j < k) ? j : k;
```

Chapter 10

1
```
switch (i) {
    case 0:
    case 2:
    case 4:
    case 6:
    case 8:
            printf("i is even\n");
        break;
    default:
            printf("i is odd (or out of range)\n");
}
```

2
```
switch (answer) {
    case 'y':
    case 'Y':
            printf("Answer is Yes\n");
            break;
    case 'n':
    case 'N':
            printf("Answer is No\n");
            break;
    default:
            printf("Illegal response\n");
}
```

Chapter 11

1
```
/*  Printing squares backwards  */

#include <stdio.h>
main()
{
  int i = 10;
  while (i > 0) {
    printf("%2d - %3d\n", i, i * i);
    i--;
  }
}
```

2   `/*   Printing squares of real numbers   */`

```
#include <stdio.h>
main()
{
    double i = 1.0;
    while (i < 2.01) {
        printf("%3.1f - %f\n", i, i * i);
        i += 0.1;
    }
}
```

3   `/*   Compound interest calculation   */`

```
#include <stdio.h>
main()
{
    int i = 0;
    float sum = 10000.0F;
    while (i++ < 25) {
        sum += sum * 0.065F;
        printf("After year %2d, sum is %9.2f\n", i, sum);
    }
}
```

Chapter 12

1   `/*   Print squares from 1.0 to 2.0   */`

```
#include <stdio.h>
main()
{
    double f;
    for (f=1.0; f < 2.01; f +=0.1)
        printf("%4.1f squared is %6.3f\n", f, f * f);
}
```

2   `/*   Sum the numbers from 1 to 10   */`

```
#include <stdio.h>
main()
{
  int i;
  int sum = 0;
  for (i=1; i <= 10; i++)
    sum += i;
  printf("Sum = %d\n", sum);
}
```

3   `/*   Sum even and odd numbers from 1 to 10   */`

```
#include <stdio.h>
main()
{
  int i;
  int even_sum = 0;
  int odd_sum = 0;
  for (i=1; i <= 10; i++) {
    if (i & 1)
      odd_sum += i;
    else
      even_sum += i;
  }
  printf("Sum of even numbers is %d\n", even_sum);
  printf("Sum of odd numbers is  %d\n", odd_sum);
}
```

Chapter 13

1   `/*   Print sum of the numbers from one to ten   */`

```
#include <stdio.h>
main()
{
  int i = 1;
  int sum = 0;
  do {
    sum += i++;
  } while (i <= 10);
  printf("Sum is %d\n", sum);
}
```

2 `/*   Calculate time to double investment   */`

```
#include <stdio.h>
main()
{
   int year = 0;
   float sum = 10000.0f;
   float target;
   target = sum * 2.0f;
   do {
      sum += sum * 0.065f;
      year++;
   } while (sum < target);
   printf("It took %d years to double\n", year);
}
```

Chapter 14

1 `/*   Raise integer to given power   */`

```
#include <stdio.h>

int main(void);
int power(int, int);

main()
{
   printf("%d\n", power(2, 8));
   return 0;
}

int power(int x, int n)
{
   int p;
   for (p=1; n>0; n--)
      p = p * x;
   return p;
}
```

2 ```
/*   Raise double to given power   */

#include <stdio.h>

int main(void);
double power(double, int);

main()
{
   printf("%f\n", power(2.0, 8));
   return 0;
}

double power(double x, int n)
{
   double p;
   for (p=1.0; n > 0; n--)
     p = p * x;
   return p;
}
```

**3**  ```
/*   Print table of squares   */

#include <stdio.h>

 int main(void);
void psquare(int);

main()
{
   int i = 1;
   while (i <= 10) {
     psquare(i);
      i++;
   }
   return 0;
}

void psquare(int n)
{
   printf("%d squared is %d\n", n, n * n);
}
```

Chapter 16

1
```
/*   Determine  screen  mode   */

#include <stdio.h>
int main(void);
main()
{
  unsigned char far *ptr;
  ptr = (unsigned char far *) 0x00400049;
  printf("Screen mode is %d\n", *ptr);
  return 0;
}
```

2
```
/*   Write character directly to the screen   */

#include <stdio.h>
int main(void);
void pch(unsigned char far *, int, int,
         unsigned char, int);
main()
{
  int i, j;
  int att = 0;
  unsigned char far *ptr;
  ptr = (unsigned char far *) 0xb8000000;
  for (i=0; i<25; i++) {
    for (j=0; j<80; j++) {
      pch(ptr, j, i, 'A', att);
      if (++att == 256)
        att = 0;
    }
  }
  return 0;
}

void pch(unsigned char far *p, int col, int row,
         unsigned char c, int att)
{
  if ((row < 0) || (row > 24))
    return;
  if ((col < 0) || (col > 79))
    return;
  p += 2 * (row * 80 + col);
  *(p++) = c;
  *p = (unsigned char) att;
}
```

```
3   /*   Print IBM character set   */

    #include <stdio.h>

    int main(void);
    void print_char(unsigned char far *, int, int,
                    unsigned char, int);

    main()
    {
      int i, j;
      unsigned char c = 0;
      unsigned char far *ptr;
      ptr = (unsigned char far *) 0xb8000000;

      for (i=0; i<25; i++)            /*   Clear screen   */
        printf("\n");

      for (i=0; i<16; i++) {
        for (j=0; j<16; j++) {
          print_char(ptr, 2 * i + 10, j+5, c, 7);
          c++;
        }
      }
      return 0;
    }

    void print_char(unsigned char far *p, int col,
                    int row, unsigned char c, int att)
    {
      if ((row < 0) || (row > 24))
        return;
      if ((col < 0) || (col > 79))
        return;
      p += 2 * (row * 80 + col);
      *(p++) = c;
      *p = (unsigned char) att;
    }
```

Chapter 17

1
```
/*   Array initialization   */

#include <stdio.h>

int main(void);

main()
{
   int data[10];
   int i;
   for (i=0; i < 10; i++)
      data[i] = i;
   return 0;
}
```

2
```
/*   Print sum of array elements   */

#include <stdio.h>

int main(void);

main()
{
   int data[10];
   int i;
   int sum = 0;
   for (i=0; i < 10; i++)
      data[i] = i;

   for (i=0; i < 10; i++) {
      printf("data[%d] = %d\n", i, data[i]);
      sum += data[i];
   }

   printf("\nSum is %d\n", sum);
   return 0;
}
```

3
```
/*   Writing a character string letter at a time   */

#include <stdio.h>

int main(void);

char name[] = "Paul";

main()
{
   int i;
   for (i=0; name[i] != '\0'; i++)
      printf("%c", name[i]);
   printf("\n");
   return 0;
}
```

4
```
/*   Writing a dot delimited string   */

#include <stdio.h>

int main(void);

char name[] = "Paul";

main()
{
   int i;
   for (i=0; name[i] != '\0'; i++) {
      if (i != 0)
         printf(".");
      printf("%c", name[i]);
   }
   printf("\n");
   return 0;
}
```

Chapter 18

1 The delimiters surrounding the file name in a #include statement indicate where the pre-processor should look for the file. Angle brackets, as in #include <stdio.h>, indicate that the file was supplied with the compiler. Double quote marks, as in #include "myfile.h", indicate a user supplied file. If a file of this name is not found in the user's file space then the files supplied with the compiler are then checked.

2 An include file usually contains function prototypes and name definitions, as in #define FRED 1. Often include files are associated with a library either supplied with the compiler or written by the user. There is no reason where an include file should not contain C code, e.g. function definitions.

3 A #define preprocessor directive does not require, and usually should not have, a terminating semicolon. It is a pre-proccessor directive, not a C statement.

4 prod3(a,b,c) ((a) * (b) * (c))

Chapter 19

1
```
/*   Summation function using one pointer   */

#include <stdio.h>
int main(void);
int sum(int *, int);

main()
{
    int k = 5;
    sum(&k, 2);
    printf("Sum = %d\n", k);
    return 0;
}
sum(int *i, int j)
{
    *i = *i + j;
    return 0;
}
```

2   `/*   Summation function without pointers   */`

```c
#include <stdio.h>

int main(void);
int sum(int, int);

main()
{
   int k = 5;
   printf("Sum = %d\n", sum(k, 2));
   return 0;
}

sum(int i, int j)
{
   return i + j;
}
```

3   `/*   String length calculation   */`

```c
int strlen(char str[])
{
   int i = 0;
   while (str[i] != '\0')
      i++;
   return i;
}
```

```c
/*   Alternative string length calculation   */

int strlen(char *str)
{
   int i;
   for (i=0; *(str++) != '\0'; i++)
      ;
   return i;
}
```

4 ```
/*   Multiplying matrices   */

#include <stdio.h>

int main(void);
void matmul(double a[2][2], double b[2][2],
            double c[2][2]);

main()
{
   double a[2][2] = {1.0, 2.0, 3.0, 4.0};
   double b[2][2] = {1.0, 2.0, 3.0, 4.0};
   double c[2][2];
   matmul(c, a, b);
   printf("%7.2f, %7.2f\n%7.2f, %7.2f\n",
           c[0][0], c[0][1], c[1][0], c[1][1]);
   return 0;
}

void matmul(double a[2][2], double b[2][2],
            double c[2][2])
{
  a[0][0] = b[0][0] * c[0][0] + b[0][1] * c[1][0];
  a[0][1] = b[0][0] * c[0][1] + b[0][1] * c[1][1];
  a[1][0] = b[1][0] * c[0][0] + b[1][1] * c[1][0];
  a[1][1] = b[1][0] * c[0][1] + b[1][1] * c[1][1];
}
```

## Chapter 20

**1**  ```
int getchar(void);
int putchar(int);
int printf(char *, int);
int scanf(char *, int*);
int sprintf(char *, char *, int);
int sscanf(char *, char *, int *);
```

2   ```
/*   Create names using preset string   */

#include <stdio.h>

int main(void);

main()
{
   int i;
   char buf[8];
   char name[] = "TMP";

   for (i=1; i<=10; i++) {
      sprintf(buf, "%s%02d", name, i);
      printf("%s\n", buf);
   }
   return 0;
}
```

**3**    ```
/*   Create names using entered string   */

#include <stdio.h>

int main(void);

main()
{
   int i;
   char buf[16];
   char name[16];

   printf("Enter base name: ");
   if (scanf("%s", name) != 1) {
      printf("Error entering data\n");
      return 1;
   }

   for (i=1; i<=10; i++) {
      sprintf(buf, "%s%02d", name, i);
      printf("%s\n", buf);
   }
   return 0;
}
```

```
4  /*   Count lower case letter frequency    */
   /*   Program assumes ASCII character set  */

   #include <stdio.h>
   int main(void);

   main()
   {
     int i;
     int counts[26];

     for (i=0; i<26; i++)           /*   Zero out counts   */
       counts[i] = 0;

     while ((i=getchar()) != EOF) {      /*   Read data    */
       if ((i >= 'a') && (i <= 'z'))
         counts[i-'a']++;
     }

     for (i=0; i < 13; i++) {         /*   Write counts   */
       printf("%c - %4d         ", i+'a', counts[i]);
       printf("%c - %4d\n", i+'a'+13, counts[i+13]);
     }
     return 0;
   }
```

Chapter 21

```
1  /*   Program to print file length    */

   #include <stdio.h>

   int main(void);

   main()
   {
     FILE *file;
     file = fopen("filename", "r");
     if (file == NULL) {
       perror("Unable to open file");
       return 1;
     }
     fseek(file, 0L, SEEK_END);
     printf("File is %ld bytes long\n", ftell(file));
     return 0;
   }
```

```
2   /*   Program to build file names   */

    #include <stdio.h>

    int main(void);

    main()
    {
       int i;
       FILE *fin, *fout;
       char name[12];
       char buf[16];

       printf("Enter root file name: ");
       if (scanf("%s", name) != 1) {
          fprintf(stderr, "Error entering root name\n");
          return 1;
       }

       for (i=0; name[i] != '\0'; i++) {
          if (name[i] == '.') {
             name[i] = '\0';
             break;
          }
       }

       sprintf(buf, "%s.in", name);
       fin = fopen(buf, "r");
       if (fin == NULL) {
          fprintf(stderr, "Unable to open %s\n", buf);
          return 2;
       }

       sprintf(buf, "%s.out", name);
       fout = fopen(buf, "w");
       if (fout == NULL) {
          fprintf(stderr, "Unable to open %s\n", buf);
          return 3;
       }

       fclose(fin);
       fclose(fout);
       return 0;
    }
```

3
```
/*   Program to count lines in a file   */

#include <stdio.h>

int main(void);

main()
{
  int i;
  FILE *fh;
  int count = 0;

  fh = fopen("fname.in", "r");
  if (fh == NULL) {
    fprintf(stderr, "Unable to open input file\n");
    return 1;
  }

  while ((i=fgetc(fh)) != EOF) {
    if (i == '\n')
      count++;
  }

  printf("Number of lines is %d\n", count);
  return 0;
}
```

4
```
/*   Program to print line lengths   */

#include <stdio.h>

int main(void);

main()
{
  int i;
  FILE *fh;
  int count = 0;
  long last = 0L;
  long this;

  fh = fopen("fname.in", "r");
  if (fh == NULL) {
    fprintf(stderr, "Unable to open input file\n");
    return 1;
  }
  while ((i=fgetc(fh)) != EOF) {
    if (i == '\n') {
      count++;
```

```
            this = ftell(fh);
            printf("Line %d has %ld characters\n",
                        count, this - last - 2L);
            last = this;
        }
    }
    return 0;
}
```

Chapter 22

1
```
/*  Pass filname as argument to main   */

#include <stdio.h>

int main(int, char **);
void exit(int);

int main(int argc, char *argv[])
{
    if (argc != 2) {
        fprintf(stderr, "Usage: %s filename\n", argv[0]);
        exit(1);
    }
    return 0;
}
```

2
```
/*   Using the argument as a file name   */

#include <stdio.h>

int main(int, char **);
void exit(int);

int main(int argc, char *argv[])
{
    FILE *fh;
    if (argc != 2) {
        fprintf(stderr, "Usage: %s filename\n", argv[0]);
        exit(1);
    }
    fh = fopen(argv[1], "r");
    if (fh == NULL) {
        fprintf(stderr, "Unable to open %s\n", argv[1]);
        exit(2);
    }
    fclose(fh);
    return 0;
```

3 ```
/*   Copy given file to the screen   */

#include <stdio.h>

int main(int, char **);
void exit(int);

int main(int argc, char *argv[])
{
   int c;
   FILE *fh;
   if (argc != 2) {
     fprintf(stderr, "Usage: %s filename\n", argv[0]);
     exit(1);
   }
   fh = fopen(argv[1], "r");
   if (fh == NULL) {
     fprintf(stderr, "Unable to open %s\n", argv[1]);
     exit(2);
   }
   while ((c=fgetc(fh)) != EOF)
     fputc(c, stdout);
   fclose(fh);
   return 0;
}
```

## Chapter 23

**1**  ```
/*   Initializing and printing a structure   */

#include <stdio.h>

struct date {
   int day;
   int month;
   int year;
};

int main(void);

main()
{
   struct date d = {15, 10, 1957};
   printf("%d/%d/%d\n", d.day, d.month, d.year);
   return 0;
}
```

2   ```/*   Date input function using structures   */```

```c
#include <stdio.h>

struct date {
   int day;
   int month;
   int year;
};

int main(void);
int indate(struct date *d);

main()
{
   struct date d;
   if (indate(&d))
     printf("%d/%d/%d\n", d.day, d.month, d.year);
   return 0;
}

int indate(struct date *d)
{
   int day, month, year;
   int valid_days[12] = {
                    31,29,31,30,31,30,31,31,30,31,30,31};
   printf("Enter date as dd/mm/yyyy: ");
   if (scanf("%d/%d/%d", &day, &month, &year) != 3) {
     printf("Invalid date format\n");
     return 0;
   }
   if ((month < 1) || (month > 12)) {
     printf("Month value is illegal\n");
     return 0;
   }
   if ((day < 1) || (day > valid_days[month-1])) {
     printf("Day value is illegal\n");
     return 0;
   }
   d->day = day;
   d->month = month;
   d->year = year;
   return 1;
}
```

3   `/*   Date output function   */`

```
void outdate(struct date *d)
{
    char *months[] = {"January", "February", "March",
                      "April", "May", "June", "July",
                      "August", "September", "October",
                      "November", "December"};
    switch (d->day) {
      case 1:
      case 21:
      case 31:
                printf("%dst ", d->day);
              break;
      case 2:
      case 22:
              printf("%dnd ", d->day);
              break;
      case 3:
      case 23:
              printf("%drd ", d->day);
              break;
      default:
              printf("%dth ", d->day);
              break;
    }
    printf("%s %d\n", months[d->month-1], d->year);
}
```

4   `/*   Determine day of week for given date   */`

```
int tellday(struct date *d)
{
    int i;
    int year;
    long ndays;
    char *days[] = {"Sunday", "Monday", "Tuesday",
                    "Wednesday", "Thursday", "Friday",
                    "Saturday"};
    int valid_days[12] = {
                    31,28,31,30,31,30,31,31,30,31,30,31};

    year = d->year - 1;
    if (year < 0) {
      printf("Unable to cope with B.C. dates!\n");
      return 0;
    }
    ndays = (long) year * 365L + (long) (year / 4 -
                    year / 100 + year / 400);
```

```
    if (d->month > 2) {
      if ((d->year % 4) == 0) {
        if (((d->year % 100)!=0)||((d->year % 400)==0))
          ndays++;
      }
    }
    for (i=1; i < d->month; i++)
      ndays += (long) valid_days[i-1];
    ndays += (long) d->day;

    i = (int) (ndays % 7L);
    printf("%s\n", days[i]);
    return 1;
}
```

Chapter 24

1
```
/*   Using unions   */

#include <stdio.h>

union fred {
   int x;
   int y;
   float z;
};

int main(void);

int main(void)
{
   union fred a;
   a.z = 4.2F;
   printf("x = %d, y = %d, z = %f\n", a.x, a.y, a.z);
   a.x = 42;
   printf("x = %d, y = %d, z = %f\n", a.x, a.y, a.z);
   return 0;
}
```

2   `/*   Printing union size   */`

```
#include <stdio.h>
union fred {
  int x;
  int y;
  float z;
};
int main(void);

int main(void)
{
  printf("Size = %d\n", (int) sizeof(union fred));
  return 0;
}
```

3   `/*   Using unions   */`

```
#include <stdio.h>

struct WORDREGS {
                unsigned int ax;
                unsigned int bx;
                unsigned int cx;
                unsigned int dx;
              };
struct BYTEREGS {
                unsigned char al, ah;
                unsigned char bl, bh;
                unsigned char cl, ch;
                unsigned char dl, dh;
              };
union REGS {
              struct WORDREGS x;
              struct BYTEREGS h;
            };
int main(void);

int main(void)
{
  union REGS regs;
  regs.x.ax = 0x1234;
  printf("ah = %02X, al = %02X, ax = %04X\n",
          regs.h.ah, regs.h.al, regs.x.ax);
  regs.h.ah = 0x77;
  printf("ah = %02X, al = %02X, ax = %04X\n",
          regs.h.ah, regs.h.al, regs.x.ax);
  return 0;
}
```

Chapter 25

1 `/*   Allocate and use integer array   */`

```c
#include <stdio.h>
#include <stdlib.h>

int main(void);
void exit(int);

int main(void)
{
   int *ptr;
   int sum = 0;
   int i, num = 250;
   ptr = malloc(num * sizeof(int));
   if (ptr == NULL) {
      fprintf(stderr, "Unable to allocate memory\n");
      exit(1);
   }
   for (i=0; i<num; i++) {
      ptr[i] = i;
      sum += ptr[i];
   }
   printf("Sum = %d\n", sum);
   free(ptr);
   return 0;
}
```

2 `/*   Determine maximum memory allocation   */`

```c
#include <stdio.h>
#include <stdlib.h>

int main(void);

int main(void)
{
   int *ptr;
   int sum = 0;
   while ((ptr = malloc(250 * sizeof(int))) != NULL)
      sum++;
   printf("Number of blocks allocated = %d\n", sum);
   return 0;
}
```

Chapter 26

1
```c
/*   Improved line length calculation   */

#include <stdio.h>
#include <string.h>
int main(void);

int main(void)
{
  int i = 0;
  char line[100];
  while (fgets(line, 100,stdin) != NULL) {
    i += strlen(line);
    if (strchr(line, '\n') != NULL) {
      if (--i == 0)
        break;
      printf("Length = %d\n", i);
      i = 0;
    }
  }
  return 0;
}
```

2
```c
/*   Ask yes or no type question   */
int ask(char *text)
{
  int i;
  do {
    printf("%s? ", text);          /*   Write question      */
    i = getchar();                 /*   Get first letter    */
    while (getchar() != '\n')      /*   Skip rest of line   */
      ;
    switch (i) {                   /*   Check reply         */
      case 'Y':
      case 'y':
              i = 1;
              break;
      case 'N':
      case 'n':
              i = 0;
              break;
      default:
              printf("Answer 'yes' or 'no'\n");
              i = -1;
    }
  } while (i < 0);    /*   Repeat till valid answer   */
  return i;
}
```

Chapter 27

1 ```
#include <stdio.h>

int main(void);
void moves(char, char, char, int);
void mover(char, char);

int main(void)
{
   moves('A', 'C', 'B', 4);   /* Move 4 rings from A
to C   */
   return 0;
}

void moves(char source, char dest, char temp, int num)
{
   if (num != 1) {   /* More than one ring   */
     moves(source, temp, dest, num-1);
     mover(source, dest);
     moves(temp, dest, source, num-1);
   }
   else
     mover(source, dest);
}

void mover(char source, char dest)
{
   printf("Move ring from %c to %c\n", source, dest);
}
```
```

Chapter 28

1
```
/*   Sorting words using a binary tree   */

#include <stdio.h>
#include <string.h>

struct item {
  char word[16];
  int count;
  struct item *left;
  struct item *right;
};

int main(void);
void bprint(struct item *);

int main(void)
{
  struct item data[50];
  struct item *first = NULL;
  struct item *p, **ptr;
  char word[16];
  int in = 0;
  int i;

  /*   Loop for all data values   */

  while (scanf("%s", word) == 1) {

  /*   Locate correct position in list   */

    ptr = &first;
    while ((p = *ptr) != NULL) {
      i = strcmp(p->word, word);
      if (i == 0) {
        p->count++;
        break;
      }
      if (i > 0)
        ptr = &(p->left);
      else
        ptr = &(p->right);
    }
```

```
   /*   Add if new entry   */

     if (p == NULL) {
        strcpy(data[in].word, word);
        data[in].count = 1;
        data[in].left = NULL;
        data[in].right = NULL;
        *ptr = &(data[in++]);
     }
   }

   /*   Now print them out   */

   bprint(first);
   return 0;
}

void bprint(struct item *p)
{
   if (p == NULL)
      return;
   bprint(p->left);
   printf("%3d - %s\n", p->count, p->word);
   bprint(p->right);
}
```

2
```
/*   Dynamically allocating structures   */
/*   P. Jarvis                08/05/1993   */

#include <stdio.h>
#include <stdlib.h>
#include <string.h>

struct item {
   char word[16];
   int count;
   struct item *left;
   struct item *right;
};

int main(void);
void bprint(struct item *);
```

```
int main(void)
{
   struct item *first = NULL;
   struct item *p, **ptr;
   char word[16];
   int i;

   /*  Loop for all data values  */

   while (scanf("%s", word) == 1) {

   /*  Locate correct position in list  */

      ptr = &first;
      while ((p = *ptr) != NULL) {
        i = strcmp(p->word, word);
        if (i == 0) {
           p->count++;
           break;
        }
        if (i > 0)
           ptr = &(p->left);
        else
           ptr = &(p->right);
      }

   /*  Add if new entry  */

      if (p == NULL) {
        p = malloc(sizeof(struct item));
        if (p == NULL) {
           fprintf(stderr, "Malloc failed\n");
           break;
        }
        strcpy(p->word, word);
        p->count = 1;
        p->left = NULL;
        p->right = NULL;
        *ptr = p;
      }
   }

   /*  Now print them out  */

   bprint(first);
   return 0;
}
```

```
void bprint(struct item *p)
{
  if (p == NULL)
    return;
  bprint(p->left);
  printf("%3d - %s\n", p->count, p->word);
  bprint(p->right);
}
```

Chapter 29

1 `/*  Getting the screen mode using the BIOS  */`

```
#include <stdio.h>
#include <dos.h>

int main(void);

int main(void)
{
  union REGS regs;
  regs.h.ah = 0x0F;
  int86(0x10, &regs, &regs);
  printf("Screen mode is %d\n", regs.h.al);
  return 0;
}
```

2 `/*  Getting the current time using the BIOS  */`

```
#include <stdio.h>
#include <dos.h>

int main(void);

int main(void)
{
  union REGS regs;
  regs.h.ah = 0x2C;
  intdos(&regs, &regs);
  printf("The time is %d:%02d:%02d\n", regs.h.ch,
          regs.h.cl, regs.h.dh);
  return 0;
}
```

3   `/*   Getting the current time using the BIOS   */`

```
#include <stdio.h>
#include <dos.h>

int main(void);

int main(void)
{
  union REGS regs;
  regs.h.ah = 0x2C;
  intdos(&regs, &regs);
  if (regs.h.ch >= 12)
    printf("The time is %d:%02d p.m.\n",
             (regs.h.ch == 12) ? 12 : regs.h.ch - 12,
               regs.h.cl);
  else
    printf("The time is %d:%02d a.m.\n",
             regs.h.ch, regs.h.cl);
  return 0;
}
```

Index

Data disk

The image processing project, and the Tagged Image File Format project, both require data. This data can be obtained free over the Internet, as explained in the project introduction (Chapter 30). If you do not have Internet access then the data can be supplied on disk by completing the following form and sending it, together with the required remittance, to the following address:

Paul Jarvis,
Imperial College Computer Sales,
Level 4, Mechanical Engineering Building,
Imperial College of Science, Technology, & Medicine,
Exhibition Road,
LONDON,
SW7 2BX

Please note that the disk contains data only, no programs are supplied. The cost of the disk is £1.50p for United Kingdom addresses, and £4.00 overseas. Please remit using a cheque (or money order) in Pounds Sterling and made payable to 'Imperial College'.

Please supply one data disk with the following format (please tick)

3.5" High density ☐   5.25" High density ☐

Name:

Address:

I enclose a cheque made payable to 'Imperial College' to the value of:

Please note that the data is supplied in good faith, but the author, Imperial College, and Oxford University Press, do not supply any warranty.